CAMBRIDGE LIBRAR

Books of enduring sch

T0229917

Life Sciences

Until the nineteenth century, the various subjects now known as the life sciences were regarded either as arcane studies which had little impact on ordinary daily life, or as a genteel hobby for the leisured classes. The increasing academic rigour and systematisation brought to the study of botany, zoology and other disciplines, and their adoption in university curricula, are reflected in the books reissued in this series.

Memorials of Sir C.J.F. Bunbury

Sir Charles James Fox Bunbury (1809–86), the distinguished botanist and geologist, corresponded regularly with Lyell, Horner, Darwin and Hooker among others, and helped them in identifying botanical fossils. He was active in the scientific societies of his time, becoming a Fellow of the Royal Society in 1851. This nine-volume edition of his letters and diaries was published privately by his wife Frances Horner and her sister Katherine Lyell between 1890 and 1893. His copious journal and letters give an unparalleled view of the scientific and cultural society of Victorian England, and of the impact of Darwin's theories on his contemporaries. Volume 5 covers the years 1865–8, and shows Bunbury's wide reading, from Latin poetry and military history to new books such as the latest edition of Lyell's *Elements of Geology*, as well as an interesting correspondence between Bunbury, Lyell, Hooker and Darwin about Agassiz's theories on glaciation.

Cambridge University Press has long been a pioneer in the reissuing of out-of-print titles from its own backlist, producing digital reprints of books that are still sought after by scholars and students but could not be reprinted economically using traditional technology. The Cambridge Library Collection extends this activity to a wider range of books which are still of importance to researchers and professionals, either for the source material they contain, or as landmarks in the history of their academic discipline.

Drawing from the world-renowned collections in the Cambridge University Library, and guided by the advice of experts in each subject area, Cambridge University Press is using state-of-the-art scanning machines in its own Printing House to capture the content of each book selected for inclusion. The files are processed to give a consistently clear, crisp image, and the books finished to the high quality standard for which the Press is recognised around the world. The latest print-on-demand technology ensures that the books will remain available indefinitely, and that orders for single or multiple copies can quickly be supplied.

The Cambridge Library Collection will bring back to life books of enduring scholarly value (including out-of-copyright works originally issued by other publishers) across a wide range of disciplines in the humanities and social sciences and in science and technology.

Memorials of Sir C.J.F. Bunbury

VOLUME 5: LATER LIFE PART 1

EDITED BY
FRANCES HORNER BUNBURY
AND KATHARINE HORNER LYELL

CAMBRIDGE UNIVERSITY PRESS

Cambridge, New York, Melbourne, Madrid, Cape Town,
Singapore, São Paolo, Delhi, Tokyo, Mexico City

Published in the United States of America by Cambridge University Press, New York

www.cambridge.org
Information on this title: www.cambridge.org/9781108041164

© in this compilation Cambridge University Press 2011

This edition first published 1891
This digitally printed version 2011

ISBN 978-1-108-04116-4 Paperback

MEMORIALS

OF

𝕾𝖎𝖗 𝕮. 𝕵. 𝕱. 𝕭𝖚𝖓𝖇𝖚𝖗𝖞, 𝕭𝖆𝖗𝖙.

EDITED BY HIS WIFE.

THE SCIENTIFIC PARTS OF THE WORK REVISED BY
HER SISTER, MRS. LYELL.

LATER LIFE

Vol. I.

MILDENHALL:
PRINTED BY S. R. SIMPSON, MILL STREET.
MDCCCXCI.

1865.

The daily journal of this year is very full, but I have omitted a great deal of it; as I have also done in 1864; which relates:—

1. To business, with which he was engaged for an hour or two almost every day.
2. To the education of his nephews, especially Clement.
3. To his charitable works.
4. To his Sunday occupations.

Our Sundays were usually spent by going to morning church, or (if it was cold), reading prayers at home, the afternoons in taking a walk; and in the evenings he had prayers and read a sermon to the servants.

[F. J. BUNBURY.]

JOURNAL.

A fine bright morning.   1865.

Read the Articles on Machiavelli in Macaulay's Essays :—very brilliant and instructive.

Read chapter 35 of Merivale's "History of the Roman Empire." The events related in this chapter (which comes down to A. U. 742), are not very important, with the exception of the campaigns against the unfortunate Alpine tribes of Rhætia and Noricum ; they chiefly concern the relations and connexions of Augustus. The conspicuous importance which the Emperor's family now begins to assume, is a new sign of the final extinction of republicanism. The most interesting things contained in this· chapter are the remarks on the characters of Agrippa and Mæcenas, the two men to whom Augustus was so incalculably indebted for their help in founding the empire. Perhaps, with all his own ability, judgment and discretion, the enterprise would have been hardly possible but for these two men. The chapter concludes with the death of Agrippa, and Merivale justly observes that,—"it is with reluctance we let the curtain drop " upon a man so eminent in public life, yet so much

1865. "less known to us than from his public career
"he deserves. There is no statesman of the best
"known period of Roman history who filled so
"large a space in the eyes of his countrymen, with
"whom we are so little personally acquainted as
"with Agrippa."

January 3rd.

The servants' dance and merry-making.

January 4th.

Read chapter 36 of Merivale. It comes down
from the death of Agrippa (A. U. 742) to that
of Mæcenas, five years afterwards ; and is chiefly
occupied with the early career of Drusus and
Tiberius. It is not very interesting. The premature death of Drusus gave him the same sort
of place in Roman history, which Edward the Sixth
or Prince Henry held in ours. There is nothing
either morally great, or picturesque, or agreeable, in
the German wars of these five years ; they are
unsatisfactory to read of, and useless to remember.
The character of Mæcenas is less grand and
dignified than that of Agrippa, but it is in most
respects very attractive. It is not surprising that
Augustus should have felt the loss of these two
friends to be quite irreparable, and that after their
death, he should not have known whither to turn for
help and comfort in trouble. Virgil died before
Mæcenas ; Horace either very shortly before, or
very shortly after ; and with these there seems to be

an end to all that is brilliant or interesting in the 1865.
reign of Augustus.

———

Dear Katharine and Harry and the children left
us. We went to Mildenhall on a visit to Lady and
Miss Lister Kaye, Sir John* being accidentally
absent ; they were very hospitable and kind. Lord
Arthur Hervey and Sarah were also staying with
them.

———

Still at Mildenhall. Very stormy and cold
weather. Fanny and all the rest went to church.
I stayed at home to nurse my cold. Read chapter
37 of Merivale: it is occupied chiefly with the
personal history of Augustus in his latter years,
and of his family ; including the disgrace and
banishment of Julia, the death of her two sons,
Caius and Lucius, who were the especial favourites
of their grandfather ; Tiberius's retirement to
Rhodes, his recall and adoption by Augustus ; and
also the conspiracy of Cinna. Poor Julia's mis-
deeds were very probably exaggerated by the prevailing love of scandal, and also by the ill will of the
partisans of Livia and the Claudii ; and yet it is
true, as Merivale says, that in such a state of
society, a man or woman who had once lost their
own self-respect, could have little or nothing to
restrain them from any excesses. It is difficult to

———

* Mildenhall Manor House and the shooting had been let to Sir John Lister
Kaye, Bart.

1865. say whether Augustus's extravagant severity towards
her is to be considered as intemperate anger, or
as part of his systematic affectation of antique
severity.

Certainly, the family relations of Augustus could
not have been very comfortable, amidst all the
jealousies and suspicions of his daughter and his
favourite grandsons on the one side, and of his wife
and *her* son on the other. Still less could the
position of Tiberius be comfortable or satisfactory.

We had a pleasant social evening at Mildenhall.

_____

Saturday, 7th January.

We returned home to Barton.

_____

Monday, 9th January.

We dined at Mr. Borton's, and met Lady Cullum,
the Abrahams, and Henry Blakes.

_____

Tuesday, 10th January.

We travelled to Sir John Boileau's, at Ketting-
ham. The Lombes dined there.

_____

Ketteringham,
Wednesday, 11th January.

Visited Wymondham Church with Sir John and
his daughters. A very fine and curious church.
The Miss Boileaus very agreeable. Mr. and Mrs.
Gurdon came to stay. Sir Archdale and Lady
Wilson dined here.

Ketteringham,
12th January.

1865.

Spent the day very agreeably with much pleasant talk.

General Boileau and Mr. Napier, the clergyman from Holkham, arrived (the last son of the late editor of the *Edinburgh Review*).

Ketteringham,
Friday, 13th January.

Took a walk with Sir John ; much agreeable talk with the Miss Boileaus. Mr. King (the geologist) and Mrs. King came to dinner.

Saturday, 14th January.

We left our kind and pleasant friends at Ketteringham, and drove to the Evans Lombes at Melton.

Melton Hall,
Sunday, January 15th.

Finished 38th chapter of Merivale ;—the last ten years of the reign and life of Augustus, from A. U. 747 to A.U. 757—A. D. 4 to 14 ;—including the banishment of the younger Julia, Agrippa's daughter, and of her brother Agrippa Posthumus, and the obscure political intrigues of the time ; the banishment of Ovid ; the campaigns of Tiberius in Germany and Pannonia ; the destruction of Varus and his legions ; and the last days of Augustus himself. Merivale discusses the cause of Ovid's banishment — one of the insoluble problems of

1865. history—ranking with the "Man in the Iron Mask,"
and the authorship of Junius's letters. He thinks
the truth probably was, that Ovid had, perhaps
accidentally or unawares, got mixed up in some
of these obscure and little understood political plots,
which somehow seem to have been connected with
the banishment of the younger Julia. This is
perhaps the most probable of all the explanations
that have been suggested ; at any rate it would
be absurd to believe the pretext of regard for morals,
which Augustus himself seems to have put forward ;
and none of the other reasons that have been
suggested appear to be tenable. Whatever the
real reason, the way in which it was done shows
how thoroughly, in those latter years of Augustus,
the gloomy, secret vindictiveness of a true despotism
had succeeded to the more liberal, open, kindly
spirit, which had characterized his government when
he was influenced by the advice of Mæcenas.

The closing scenes of the life of Augustus are well
related by Merivale, and the observations on his
career are judicious.

Melton Hall,
January 16th.

Walked with Mr. Lombe to see the fine chestnut
trees and the old church. Went with Mrs. Lombe
and Fanny into Norwich, and called on Sedgewick.
General Angerstein came. A dinner party. Very
pleasant.

January 17th.   1865.

We left Melton, and returned home, travelling with General Angerstein from Wymondam to Thetford.

_____

Wednesday, January 18th.

Read in the last number of the *Geographical Society's Proceedings*, the report of Captain Burton's paper on "Lake Tanganyika" (really on the sources of the Nile in opposition to Captain Speke), and of the debate that ensued on it. There really seems much room for doubt whether the true head waters of the Nile have yet been discovered.

We dined with Mr. Beckford Bevan.

_____

January 19th.

Heard of the death of Mr. Charles Greville. We dined at the Benyons at Culford.

_____

January 20th.

Began to read Montalembert's " De l'avenir politique de l'Angleterre." We arranged the newly-bound books, and went on with the catalogue.

_____

Saturday, 21st January.

Finished chapter 39 of Merivale's "Roman Empire." It contains a review, as far as the existing materials will admit, of the statistics of the Empire in Augustus's reign ; an account of the operations of the census, and of the map or "picture" of the

1865. Empire, which was executed under the direction of Agrippa;—an attempt to estimate (very vaguely as he admits), the population of Italy.

———

Read chapter 2 of St. John's Gospel with Alford's Notes. Read prayers with Fanny. Very cold.

———

Finished chapter 40 of Merivale. This begins with notices of the principal cities of the Roman Empire; and much the greater part of it is occupied by a very copious, clear and interesting description of Rome itself, as it appears to have been in the time of Augustus. Merivale enters carefully into the question of population of ancient Rome, and concludes, seemingly on good grounds, that the population within the Servian walls can hardly at any time have exceeded 420,000; and that even including the suburbs, it could hardly in the reign of Augustus, amount to 700,000.

Cecil and Clement arrived.

———

Very cold.

We dined (with Lady Cullum) at Hardwick. Met Mr. and Mrs. Waterton, the Abrahams, Wilsons, &c.

———

Wednesday, 25th January. 1865.

Very cold.

Meeting of Council of Suffolk Rifle Association at "The Angel." Maitland Wilson, Barnardiston, Tyrell, Colonel Anstruther, Major Parker.

———

Thursday, 26th, January.

Snow.

We went to General Angerstein's, at Weeting. Found Lady Wallscourt and Miss Blake at Weeting. Mr. and Mrs. Sparke came to dinner.

———

Weeting,
Friday, 27th January.

Desperate weather—blowing hard with furious sleet and snow. Cecil and Clement went shooting with the General. We stayed within doors all day.

Beautiful pictures at Weeting, especially Sir Joshua's. Garrick between "Tragedy and Comedy." "The Nymph," and several fine portraits of the Angerstein family.

———

Saturday, 28th January.

A fine bright sunny day. The ground covered with snow. We walked about the grounds with Miss Blake,* she is very pretty and pleasant. After luncheon we returned home.

———

Sunday, 29th January.

Read prayers with Fanny.

Finished reading chapter 41 of Merivale. (The

* She afterwards married General Arthur Upton.

1865. last of the 4th volume) of his history of "The Romans under the Empire."

This chapter contains a view of the manners and social state of the Romans under Augustus :— their occupations and amusements : the theatres, the circus, the amphitheatre, the baths, the feasts, &c., well put together and agreeably written, but not including anything particularly new to me, except which relates to the rhetorical exercises with which I was not acquainted, or rather I did not appreciate the important place they held in the pursuits of a Roman gentleman. I am wrong— another thing quite new to me was the amusing extract from Petronius, giving us (in a parody) an idea of the published notices which served the purpose of newspapers to the people of Rome.

The remarks on the poets of the Augustan age, especially in Virgil and Horace, which fill the latter part of this chapter, are interesting, and I think judicious ; in particular, the observations on the political tendency and object which these two great poets had in view in much of their writings, are very curious.

LETTER.

Barton,
January 30th, 1865.

My Dear Mary,

My copy of the " Elements " arrived just before we set out for General Angerstein's, and though very much pleased to see it, I deferred till

our return, both the beginning to read it and the 1865.
writing my acknowledgments. Now I can begin
upon it with full satisfaction, and have done so,
having to day read the 10th chapter, on " The
Recent Period and the Antiquity of Man," which I
much approve. Pray give my best thanks to
Charles Lyell for the book. I mean to read carefully through all the chapters on the tertiary and
post-pliocene formations, and to pick out from the
remainder all that appears to be new, and when I
have finished I will write him my observations.
The comparative table of formations at the
beginning appears exceedingly good and useful.

I was extremely glad to hear of your safe return
from Berlin, and had intended to write to welcome
you; but I *sometimes* delay to execute my good
resolutions in the way of letter writing. I hope you
are both of you really the better for your tour.

We have winter in good earnest at last, and deep
snow, to the great delight of Scott and the farmers;
and indeed it will be a great benefit to the country
by supplying the springs and ponds.

I have been very well since our delightful week in
Norfolk with the Boileaus and Lombes, which
quite set me up.

Our two days with General Angerstein were also
very pleasant. He was extremely courteous and
hospitable, and very agreeable; clever and rather
odd, and therefore the more entertaining. His
cousins, Lady Wallscourt and Miss Blake, who did
the honours as ladies of the house for the time
being were also very agreeable. His house is a

1865. very good one, and rich in fine pictures of Sir
Joshua Reynolds and Sir Thomas Lawrence. By
the way, when is Mr. Taylor going to bring out
his " Life of Sir Joshua ? " Do you hear anything
of it ?

Cecil and Clement are with us ; they are fine
young men.

With much love to Charles Lyell and to
Katharine,

I am ever your very affectionate Brother,
CHARLES J. F. BUNBURY.

JOURNAL.

Wednesday, 1st February.

A very fine day. Snow nearly gone. Lady
Cullum, Miss Lister Kaye and Mr. and Mrs.
Barrington Mills came to us.

Thursday, February 2nd.

A very mild day. Walked round the park and
plantations.

Finished chapter 42 of " Merivale's Roman
History," the beginning of the reign of Tiberius,
and the campaigns of Germanicus in Northern
Germany. I am quite disposed to believe with
Merivale that Tacitus has greatly exaggerated the
victories of Germanicus; it seems certain at any
rate that they produced no important or permanent
effects. As to the character of Germanicus himself,
it seems probable that his immense popularity in

1865.

Rome was mainly owing to the contrast of his manners with those of the melancholy and sullen Tiberius, in addition to the usual tendency to admire and hope everything from a promising young prince. It would be in the natural order of things that the Emperor should be jealous of him ; but one can hardly say that Tiberius was wrong as a matter of policy in discontinuing the attempt to conquer Germany.

The Arthur Herveys and Mr. Abraham came to us. Mr. and Mrs. Waterton dined with us.

Friday, 3rd February.

Mild, but damp. The Arthur Herveys went away. Sir John and Lady Lister Kaye, Mrs. Abraham and Mr. Lott arrived. A very large dinner party.

4th February.

My 56th birthday. *Deo gratias.* Cecil went away, and afterwards all our guests.

LETTER.

Barton,
February 5th, '65

My Dear Katharine,

I thank you very heartily for your kind and pleasant letter which I received yesterday, and for your good wishes on the occasion of my birthday. The "flying years" (as Horace says), are indeed slipping away ; but I do not feel inclined, like

B

1865. him, to repine at their flight; I have every reason to feel thankfulness for the past and present, and humble hope and trust for the future; and it would be ungrateful indeed to be dissatisfied, because one may not have the advantages of youth combined with those of mature age. It does not require "birthday presents" to assure me of your kind feeling and thought of me: but I thank you for the little chain, which will be very convenient, and would serve to remind me of you, if that were necessary. Pray thank darling Rosamond for her sweet letter to me; say that I like it very much, and that I will write to her as soon as I have time. We are now, for a wonder, *almost* alone,—no one with us but Clement, Cecil having left us yesterday morning to return to Brighton. Our gaieties last week went off very well; the Arthur Herveys and the Abrahams are always very agreeable, Lady Cullum always both amusing and genial, and full of kindliness; and Miss Lister-Kaye is a most amiable person. Sir John and Miladi could not come till Friday, in consequence of his having an attack of bronchitis, and I began to think it was fated that he and I should *never* meet; but he did come at last, and we got on very tolerably well together. He is not exactly a man I should ever be intimate with, but he is a gentleman, and he is certainly a first-rate tenant. I agree with you that giving and receiving hospitality is a very great pleasure, but I find it terribly destructive to anything like study. Now we have a prospect of eight or nine days of quiet, and I hope to get some reading. I am going, with great pleasure, through all the new parts of the new edition of "The Elements," (which are by no means elementary) ; and likewise going on with Merivale, which is a *pièce de résistance.* I was sorry to see Dr. Falconer's death in the papers : not that he was a man to whom I ever felt any very strong attraction, but I think he must be a loss to science.

Nothing can be more true than what you say of Fanny—that she is always thinking of making others happy, and labouring to do it. She is like her dear Father and Mother in this. Her picture has been very much admired by our guests. Pray thank dear Mary for me, for her kind letter, which I will answer soon.

With much love to your husband and the dear children,

> I am ever,
> Your very affectionate Brother,
> C. J. F. BUNBURY.

Do you know this *very naughty* riddle ?
Why is marriage like the letter T ?
Because it is the end of quiet and the beginning of trouble.

JOURNAL.

Sunday, 5th February.

Read prayers with Fanny, and Robertson's beautiful Sermon on the New Commandment of Love.

Monday, 6th February.

A young tree, a Cephalonian Fir was planted for Clement.

═══

LETTER.

Barton,
February 7th, 65.

My Dear Mary,

I thank you very much indeed for your kind letter and good wishes on my birthday. I do indeed feel a most entire belief and confidence in your love for me, which as you say is now of long standing, and which, I am sure you know, is fully reciprocated by me ; it is very pleasant to me to receive from time to time fresh assurances of it. I do not at all require " birthday presents " to confirm that belief, nor do I expect them : though of course I am thankful for them when they come.

Fanny seems pretty well to-day. Now that we are alone, I hope to make more progress with " The Elements." I yesterday read through the 15th chapter, on the " Flora of the Miocene Period," with great satisfaction and approbation. I am glad that you persuade Lyell to go out sometimes. I hope he is a tractable patient, as I am when I have anything the matter with me. Did you observe it was said in the papers, that Mr. Strutt is the first heir to a peerage who ever has been Senior Wrangler ? This is *literally* true, but not in substance. The present Duke of Devonshire was

second Wrangler, but he was *first* "Smith's Prize" 1865.
man, which is as high an honour as being Senior
Wrangler.

With very much love to Lyell,

I am ever your very affectionate Brother,

CHARLES J. F. BUNBURY.

JOURNAL.

Tuesday, 7th February.

Wet weather. Clement went away. Had a kind
letter from dear Cissy. Finished chapter 43rd of
Merivale, containing chiefly the death of Germanicus, the enthusiasm of sorrow with which his
funeral was celebrated at Rome, the trial and
suicide of his enemy Piso. We here see the shades
gradually thickening and darkening over the
character of Tiberius ; his unpopularity increasing,
and his faults developing themselves, thoroughly
unamiable he no doubt was, but I think with
Merivale, that it is impossible to believe that he
contributed to the death of Germanicus ; nor does
there seem to be at all a strong case, in this matter,
even against Piso. Went on with "The Chan-
cellors" with Fanny.

LETTER.

Barton,
February 9th, 1865.

My Dear Joanna,

I was actually thinking of writing to you
when I received your kind and very pleasant letter,
for which I thank you heartily. I am very glad

1865. that you have a young botanical friend, and hope you will pursue the study with him. Knowing something of Italy, I can believe it possible that there may be as many wild plants in flower as he says, though it is difficult to *realize* it when we are in the midst of frost and snow. Yes, the ground this morning was quite white with snow. Not that we are quite without signs of approaching spring. The yellow Aconites are in bloom, and the Snow-drops beginning to peep and the leaves of the white Saxifrage coming up, but still we are very wintry. It is in April and May that Italy is a real paradise of flowers, and then I daresay you will feel yourself impelled to study botany : the more as you will have a real live professor at hand to refer to in case of difficulty. I have received a copy of the new edition of Lyell's "Elements," and have read through the chapters on the recent post Pliocene and tertiary periods which contain the greatest amount of novelty. They are extremely well-done, especially that on the Miocene Flora. I am now in the 5th volume of Merivale in the reign of Tiberius. I think he makes out with great skill the character of that Emperor—not a monster, but a man of a narrow mind and a morose, gloomy temper with a great tendency to morbid melancholy and suspicion : distrusting himself and suspecting every one else ; the circumstances of his position fostered his faults into crime.

But the interest of the history goes off very much indeed after the time of Augustus. Now that we are alone, I am reading to Fanny in the evenings some of Lord Campbell's "Lives of the Chan- cellors," which are very entertaining : we have had a great deal of pleasant society since I wrote to Susan. We spent a delightful week in Norfolk in the middle of January : first four days with the Boileaus whom I liked even more than when they were here in 1863.

We have also had some pleasant company here, and several dinner parties in the neighbourhood. Cecil and Clement have lately left us : they are very fine and gentleman-like young men. Mr. Eddis's drawing of Fanny is beautiful and has been much admired by our visitors. I hope you will see it one day.

(February 11*th.)* More snow has fallen and the weather is very cold. I cannot say that I look forward with pleasure to a journey into Essex next week. We are going to pay a visit to Lord and Lady Rayleigh, whose eldest son, Mr. Strutt, is Senior Wrangler at Cambridge this year ; on the 20th we hope the Kingsleys will be here, and on the 22nd Sir Edmund and Lady Head.

We are still going on whenever we are alone with the Catalogue of books, and have been engaged this week in the dark passage, which had been made the receptacle of a great many stray books from other parts of the house. Our new Fern house seems to answer well, and the plants in it look in good health.

With much love to dear Susan, I am ever,

Your very affectionate Brother,

CHARLES J. F. BUNBURY.

JOURNAL.

1865. Read the Duke of Argyle's address to the Royal
Society of Edinburgh, containing a very able and
judicious discussion of the Darwinian theory.

We visited some cottages.

Very cold. Read prayers with Fanny, and one
of Robertson's sermons. Read parts of chapter 20,
21 and 22, of Lyell's " Elements," picking out more
particularly, the novelties : the account of the
wonderful fossil bird, the Archæoptery ; the mammals of the Purbeck beds and of the Trias; and the
interesting marine beds of St. Cassian, and of
Hallstadt, the deep-sea representatives of the
Trias.

Finished 44th chapter of Merivale. It treats of
Tiberius's administration, and in particular of the
terrible laws of treason *(læsæ majestatis)* and the
atrocious system of *delations*, or informations under
these laws. Merivale shows how this abomination
which grew to such monstrous proportions under
the Empire, originated from the practice of impeachments, which had flourished under the
Republic. Even in the best times of the Republic,
and most especially in its latter days, it had been
the practice of aspiring orators and ambitious

young men to signalize their zeal and abilities 1865. by bringing impeachments against men who had held high official situations, and especially for maintaining charges of oppression and extortion (very often well founded we may believe) against those who had governed Provinces. Under the Empire, this easily degenerated into the practice of raking up, on the most trivial grounds, charges of disrespect to the Emperor. But although Tiberius even in the early years of his reign was too ready to listen to these Delators, Merivale contends that he had on the whole great merit as an administrator. He certainly does appear to have been remarkable for industry and application to business ; as good a man of business as Philip II. of Spain ; and in a certain degree very conscientious. But, as Merivale observes, his patience, industry and discretion were disparaged by a perverse temper, a crooked policy, and an uneasy sensibility. Nevertheless there seems reason to believe that the condition of the Provinces was better under his Government than it had ever previously been.

Tuesday, 14th February.

Received Lord Derby's " Homer," and Macdougal's " Modern Warfare." We dined with Sir James Simpson ; met Lady Cullum, Mr. Abraham, the Cheeres, &c.

Wednesday, 15th February.

Very cold. We went to Lord Rayleigh's, Tarling

1865. Place, in Essex. Lord Wrottesley at Tarling. Sir Thomas and Lady Western and Miss Western came to dinner.

——— ———

Thursday, 16th February.

At Tarling. Very cold weather—ground covered with snow. Mr. Strutt lectured on "The Sun," at the Institute at Witham—very good. Lord Wrottesley pleasant.

———————

Friday, 17th February.

At Tarling. More snow.

Lord Wrottesley went away. A large dinner party. During these two days at Tarling, I read a considerable part of Macdougal's new book "Modern Warfare as influenced by Modern Artillery." The title is rather deceptive, for only a small part of the book really has any reference to the modern improvements in artillery and fire-arms, or to the modifications which they must introduce into warfare.

It is only in the second chapter that this subject is seriously discussed; the rest of the work is a kind of sequel to his "Theory of War;" treating of the general principles of tactics, and illustrating them by a full examination of a number of actual cases. All this is treated in the same clear, simple unaffected and sensible manner which was characteristic of his former book. I might perhaps be inclined to doubt whether it was worth while to borrow so large a proportion of his examples from the history of the Peninsula War; for it would seem that every

Englishman who takes any interest in military 1865. science must be familiar with that work.

Saturday, 18th February.

We left Tarling. More snow. We stopped at the Manningtree station, and drove thence to Stutton, to see our dear friends, the Mills; had luncheon with them ; thence to Bentley station and so by the railway to Thurston, and safe home. Thank God.

Sunday, 19th February.

Read prayers and one of Robertson's sermons to Fanny.

Read books 3, 4 and 5 of Lord Derby's translation of " The Iliad." Very good.

Monday, 20th February.

A long talk with Cooper on Mildenhall business— a satisfactory report of the audit.

The Kingsleys arrived. A very pleasant evening with them. Tremendous weather.

Tuesday 21st February.

Very cold.

Looked over some dried plants with Kingsley. In the afternoon we looked over some prints with them. The Arthur Herveys arrived. A very pleasant evening. Rose Kingsley and the Hervey girls sang sweetly.

Fog and thaw.

Kingsley went to Cambridge and returned in the afternoon. Took a walk with Arthur Hervey and Mr. Strutt. Sir Edmund and Lady Head, and General Angerstein arrived.

Thursday, 23rd February.

A fine and mild day. Arthur Hervey, Kingsley, Mr. Strutt and I, went into Bury, and saw the Museum,

We were all photographed by Spanton.

Lady Cullum and Mr. Abraham arrived.

Friday, 24th February.

Our dear friends the Kingsleys went away, also General Angerstein.

We showed the house to the Heads. Sir James Simpson came to dinner.

Finished chapter 45th of Merivale's "Roman Empire." It includes the years (A. U., 774, 782, A. D. 21, 29), in which the evil tendencies of Tiberius's reign became established and confirmed, the ascendancy of Sejanus, the complete triumph of the *delators*, the death of Drusus the son of Tiberius (supposed to have been poisoned by Sejanus with the aid of his own wife) and the retirement of Tiberius to Capreæ. A gloomy period.

A beautiful day. Sir Edmund and Lady Head, the Arthur Herveys, Lady Cullum and Mr. Abraham went away in the morning, and Mr. Strutt in the afternoon.

Received a present of prints from General Angerstein, and wrote to thank him.

Sunday, 26th February.

My cold very heavy. Read prayers with Fanny. Read Arthur Hervey's article on the "Book of Kings," in Smith's "Dictionary of the Bible."

LETTER.

Barton,
February 26th, '65.

My Dear Lyell,

You are quite right about the Sago, and yet Heer is not wrong. Sago is yielded by plants of two different families, Palms and Cycads. That which is of the finest quality, it is said, is prepared from the interior of the stems of certain Malayan Palms, chiefly of the genus Sagus. But Sago is prepared also to a great extent in Japan and China from the pith of Cycas revoluta, and in the Moluccas from that of Cycas circinalis. Nevertheless, it seems rather inconvenient and confusing to give the name of Sago trees as equivalent to the Cycadeæ, as it would belong more strictly to the Palms of the genus Sagus. Heer's "Urwelt" I once saw in your house, but that is all I know of it.

1865.   We have had a week of exceedingly pleasant
society : the Kingsleys, Sir Edmund and Lady
Head, the Arthur Herveys, Mr. Strutt and others.
I have been for some time intending to write to you
my remarks about your "Elements," but first our
journey into Essex, and then this company came in
the way, and now I am stupified by a heavy cold,
(caught by sitting a long time out-of-doors as a
component part of a group to be photographed) and
my head is not clear enough.   But it is of the less
consequence, as I have nothing of importance to
say except in praise ; my criticisms being few and
not of much importance.

Fanny is gone to church : I do hope she will not
catch cold, as she has kept so well hitherto during
all this company time and throughout the severe
weather.

Much love to dear Mary.

Ever your very affectionate Friend,

CHARLES J. F. BUNBURY.

JOURNAL.

Monday, 27th, February.

A very fine day.   Walked in the sunshine and
enjoyed the Snowdrops and Crocuses.

Tuesday, 28th February.

The Barton Rent Audit.   Very satisfactory.
Fanny went to Mildenhall and returned to dinner.
Went on with the catalogue of my herbarium.

Read books 7 and 8 of Lord Derby's "Homer." 1865.
The simile of the stars at the end of the 8th book
appears to me particularly well translated.

Finished chapter 46th of Merivale, which comes
down to the death of Tiberius (A. U, 790, A. D.
37). It comprises a wretched and dismal portion
of history, painful and disgusting to read of,
unredeemed by any great characters or noble deeds.

Merivale is certainly disposed to treat Tiberius
with at least sufficient indulgence : but he makes
out his character with great skill and probably with
a great deal of truth, and it is as repulsive a one as
need be. Not indeed the sort of exaggerated
monster that one imagines him from the ordinary
histories : but a man of a gloomy, morose, unhappy
temper, prone to melancholy and suspicion,
additionally soured and rendered more timid and
reserved by the inviduous and painful position in
which he had been kept so long by Augustus ; he
was at last placed in a situation of boundless
power, and at the same time of almost boundless
danger. Besieged by flatterers, and those not
merely of mean condition, but of the noblest blood
and highest dignity : urged even by the Senate to
guard his own safety as the safety of the state,
irritated by seeing the superior popularity of his
brother's family, feeling himself more and more
unequal as he grew older, to the enormous task of
personally governing such an Empire, and thence
falling more and more under the influence of
worthless favourites : all his evil qualities became
continually aggravated and intensified as he grew

1865. older, and his position made their effect tremendous. As a private man, he would have been odiously unamiable: as an Emperor, "the patron of Sejanus and of the *delators*," he was atrocious.

As to the excesses imputed to Tiberius while in his retreat at Capreæ, Merivale seems to think them mere scandal.

———

Attended Rifle Association Meeting at Bury— time wasted. Finished Macdougal's new book on " Modern Warfare and Modern Artillery." In the last chapter he returns to the professional subject of the book—the influence which the tremendously increased power of artillery is likely to have on battles. He shows how much more difficult and dangerous the operations of infantry in attack, will be rendered by this formidable artillery ; and though he proposes methods of meeting the difficulty, he does not seem to have much confidence in them. It appears clear enough as he shows, that infantry who are to advance over ground swept by such an artillery fire, must advance in open order, and at as rapid a pace as possible; but the difficulty consists in their forming line again with sufficient quickness and firmness, after such a *race* to encounter the enemy's infantry. On the whole he is of opinion that all infantry henceforth ought to be trained as light infantry, and that the utmost care ought to be taken to exercise them in such a manner as to give them the greatest possible activity of body, and quickness of intelligence. At the same time, he 1865. repeats here what he had already laid down in an early chapter, that the increased power of artillery would give a great advantage to the defenders of a position over its assailants ; and that the superior talents of a General will be shown in so conducting a campaign, as to oblige the enemy to attack him in position. He points out also very justly, that the true essential principles, both of strategy and tactics, remain unchanged, although the details of an actual battle may be modified by the recent inventions.

In chapter 10th (on " Mountain Warfare,") he gives a remarkably clear summary and explanation of the very complicated operations of the Pyrenees, in Wellington's and Soult's campaign of July, 1813. The substance of this is of course taken from Sir William Napier : but it appears to be exceedingly well done ; and the exclusion of all the *incidents*, all the circumstances of actual fighting certainly enables me to understand more easily and more clearly the rationale of the whole. We thus see more clearly what were the advantages and disadvantages on either side, what each General intended to do and how the one was baffled.

Thursday, March 2nd.

We went on with the book catalogue ; and I went on with my Herbarium Catalogue.

LETTER.

Barton,
March 2nd, 1865.

My Dear Lyell,

1865.   I have written on a separate paper, which
I inclose, the *special* remarks I have made in reading
your new edition ; but really they are of so little
importance that they seem hardly worth sending.
I have read with care and with great admiration
your chapters on the Recent, Post-pliocene and
Tertiary periods. I admire, especially the way in
which you have contrived to pack into so small a
space a clear summary of the various evidences of
the antiquity of man. The theory of the formation
of lake-basins by ice is a very difficult question, and
I dare say it could not have been treated more
clearly in so narrow a space than you have done it ;
but I must own I find it very hard to understand.
I must try again. The whole subject of the Tertiary
Flora, I think excellently well treated. I most
especially and cordially approve of the observations
on the " Alleged difference in the degree of affinity
of the Miocene plants and shells to the living
creation." The truth is there is much more vague-
ness and uncertainty about such comparisons than
many geologists are willing to allow, and this is
mainly owing to the want of any *common measure*
(as arithmeticians would say)—the want of anything
like a fixed standard or principle of specific dis-
crimination. I was delighted with another passage

in which you have touched on the same subject—on 1865.
the vacillating and arbitrary opinions of palæontol-
ogists as to species—I mean at page 214. *But*
then it does appear to me that in other parts of
your tertiary system you have relied with (what
seems) excessive confidence on these same fine-
drawn distinctions of species.

I believe the majority of conchologists and
geologists would go along with you here, so I dare
say I am wrong.

I have also read with great interest, your account
of the Aix la Chapelle plant bed in the cretaceous
system ; and that of the Rhætic or St. Cassian
beds. Only I think, it would be well to give a little
geographical information as to whereabouts St.
Cassian and Hallstadt are.

I showed your new edition to Kingsley, who had
not seen it before, and was much delighted with it.

It is very pleasant to see the flower-beds before
our windows, gay with the Crimean Snowdrops,
with Crocuses of three kinds, a beautiful blue Squill
and Hepaticas.

With much love to Mary,

I am ever your affectionate friend,

CHARLES J. F. BUNBURY.

JOURNAL.

Friday, March 3rd.

Finished the Mildenhall Estate accounts. Arranged minerals.

Saturday, 4th March.

News of the surrender of Charleston.

————

Sunday, 5th March.

Read 16th chapter of Luke, in Greek, with
Alford's notes. We went on with the catalogue of
books.

————

March 6th.

Finished chapter 48th of Merivale, coming down
to the death of Caius. I think Merivale has dwelt
with greater length than necessary on this worthless
and uninteresting reign. The character of Caius
appears much more simple and easy to understand
than that of Tiberius. By nature weak, excitable
and sensual, miserably ill educated (or rather
systematically *mis*-educated if we may coin such a
word); it seems clear that his very strong natural
tendency to madness was inflamed into actual
insanity by the exciting effect of his position,—
added perhaps to attacks (which are recorded by
these historians) of physical illness. Merivale, though
he admits Caius's tendency to madness, appears
unnecessarily sceptical as to the extravagances
imputed to him. It seems to me that they are, or
have been, men in our own time, who, under the like
circumstances, would have been quite capable of equal
freaks of tyranny. Merivale, however, very sagaciously
points out, how the peculiarly oriental style of Caius's
despotism may have originated in the influence of
his intimate associate Herod Agrippa, who was
afterwards king of Judea, and was thoroughly

imbued with the true Eastern notions of sovereignty. 1865.
After all, however detestable the tyranny both of
Tiberius and of Caius, no small share of the blame
ought to fall on the base and profligate servility of
the Roman Senate.

Tuesday, March 7th.

Mr. Abraham came to luncheon and was very
pleasant.

Thursday, March 9th.

Finished chapter 50th and volume 5 of Merivale,
ending with the death of Claudius (A. U. 807, A. D.
54). Merivale's account of that unfortunate, weak,
well-meaning, ill-used prince, is fair and discriminating. He points out many puzzling incon-
sistencies in the existing accounts, which may
certainly lead us to suspect that the historians
borrowed much from the mere scandalous chronicles
of the time. Whatever may have been the relative
demerits of Messalina and Agrippina, (Merivale is
rather inclined to apologise for the former), it would
certainly seem that the crimes and abominations
which have made the reign of Claudius so notorious,
affected chiefly those more or less connected with
the court. The empire generally appears to have
been prosperous and reasonably well governed.

Friday, March 10th.

Wrote some notes on Conifers cultivated here.
Went on with the catalogue of books.

1865.   Saturday, March 11th.

Read chapter 51st, the first of the 6th volume of Merivale.

This contains the history of the conquest of Britain, (at least of the southern part of the island) by Ostorius Scapula and Suetonius Paulinus : the heroic struggle of Caractacus, the destruction of the Druids, and the revolt of the Britons under Boadicea. It is well told. Merivale justly observes that the treatment of Caractacus by Claudius is honourably contrasted with that of Vercingetorix by Julius Cæsar.

Sunday, March 12th.

Read prayers with Fanny and read some of Robertson's Sermons.

Read Alford's commentary on the difficult chapter, Matthew 24th, and wrote some extracts from it.

Finished my notes on the Miocene fossil Ferns, and began some on the tertiary Conifers from Heer's "Flora Tertiara" and Göppert's "Fossil Coniferæ."

Tuesday, March 14th.

We went to luncheon with the Arthur Herveys and had a very pleasant talk with them.

Thursday, March 16th.

Fanny went over to Mildenhall to the school inspection and returned to dinner with Mr. Meyrick, the school inspector. Lady Cullum and the Pellews dined with us.

LETTER.

Barton,
March 16th, 1865.

My Dear Joanna,

Our last *batch* of company, from the 20th to the 25th of February, was uncommonly agreeable,—You know the Kingsleys, I think ; and you know Sir Edmund and Lady Head, whom I knew but very slightly before this, but whom we found very agreeable. Then there were the Arthur Herveys, Lady Cullum, General Angerstein, Mr. Abraham, and young John Strutt, the *Senior Wrangler* of the year, whom we like very much. All young and old appeared to the best advantage, and seemed to suit one another—to fit in together, as it were particularly well, so that I hope they as well as we found it pleasant. The only drawback was that the Kingsleys could stay so short a time, and in particular that Kingsley himself was obliged to take the greater part of one day, even out of that short time, for his lecture at Cambridge.—Since then, we have been very quiet and very happy by ourselves.—Now for books: I have lately read (rather skippingly) above half of Lord Derby's translation of the Iliad ; and I must say I think his perseverance wonderful ; that he should have had pleasure in translating many of the finer portions, I can easily understand ; but that he should have laboured steadily through

1864. all the lists of killed and wounded—through all the anatomical descriptions of wounds, does show marvellous perseverance ; as it can only have been the relaxation of his leisure, it is exceedingly close to the original, and has I think a great deal of poetical spirit ; many of the similes are very finely rendered, and I must confess I find it a little tedious on the whole ; but in that respect, it does not differ from every other Epic that I have ever read. I remember indeed that I read through Pope's Homer with great delight when I was a boy, but I doubt whether I could do so now. The Orlando Furioso I do not call an Epic—still less Walter Scott's poems or Southey's. I have not yet begun to read the Emperor Napoleon's "Julius Cæsar," though I have bought it ; but in cutting the leaves open I observe that he seems to accept without question the whole story of the Monarchy and early Republic as told by Titus Severus and Dionysius of Halicarnossus, as if he had never heard of Niebuhr, or Michelet, or Arnold. We have not made any great progress in the "Life of Lord Somers" (Lord Campbell's) for Fanny has been a little apt to fall asleep in the evening, as soon as I began reading to her !

I am in the sixth volume of Merivale, in the middle of the reign of Nero. He is a good and careful historian, and evidently knows his subject well, but I do think his book is a little longer than necessary ; the part between Augustus and Vespasian might I think have been shortened with some advantage. From Tiberius, onward, it is

a dreary, odious, almost disgusting story of crimes
and follies, hardly redeemed by any indication of
greatness or nobleness of character.

Seneca makes a very indifferent figure; and
however we may abominate the weakness of Tiberius
and Caligula and Nero (Caligula indeed might be
acquitted of insanity)—we feel that a large share of
the blame ought to fall on the base and profligate
servility of the Senate. At a rather later time they
made a considerable *rally*. The campaigns in
Britain of Ostorius Scapula and Suetonius Paulinus,
form a splendid episode; and Merivale very justly
points out that the treatment of Caractacus by
Claudius is favourably contrasted with the cruelty of
Julius Cæsar towards Vercingetorix. Indeed, poor
Claudius seems to have been a really well-meaning
man, but he was weak, and had the misfortune
to fall under the influence of two atrocious women
in succession.

We have (or rather Fanny has) negociated an
exchange of pictures with Sir Henry Blake; we
have given him his grandfather, Sir Patrick, the
lengthy gentleman in a red coat, whom you may
perhaps remember in our entrance hall; and have
obtained in exchange Lady Blake, also by Sir
J. Reynolds; a much more beautiful picture, but in
a deplorable condition. I hope we shall be able to
get her well restored in London. The Library here
is to be painted while we are in London—an awful
operation; I wish it were well over.

I rather think you did not see our *bird table* before
you left England; it was an operation in the

1865. summer. All this winter it has been much in vogue,
and I think we shall be able to boast of having the
best-fed Sparrows in the country. But besides
Sparrows, we have four kinds of Titmouse, most
amusing little creatures ; also Blackbirds, Red-
breasts, Chaffinches, and Nut-hatches.

With much love to dear Susan, believe me ever,
dear Joanna,

Your very affectionate Brother,

CHARLES J. F. BUNBURY.

JOURNAL.

Saturday, 18th March.

Went to Stowlangtoft and had a pleasant talk
with the Rickardses.

Monday, 20th March.

Finished chapter 53rd of Merivale ; coming down
to A.U. 819, the 12th year of Nero's reign, when his
wickedness was at its very height. Merivale makes
no attempt to palliate or explain away the atrocities
of Nero ; one sees clearly from his narrative why it
is that this Emperor has generally been allowed
such a pre-eminence of infamy. He may certainly
be considered as a type of the worst kind of tyrant ;
not only because of his monstrous and horrible
profligacy, but also because his enormous cruelties
all sprang from sheer cowardice. No doubt the
anger of the Romans—at least of the Nobles—
against him was inflamed by their haughty and

arrogant contempt for those artistic tastes which he 1865.
indulged to such excess; and so far one cannot
much sympathize with them; but after making
ample deductions on this account, there remains
fully enough to justify the detestation in which his
name has been held for so many centuries.

Thursday, March 23rd.

The assizes at Bury. I was foreman of the grand
jury. Dined with the Judges.

Friday, March 24th.

Read some geological papers, and read the
remainder of chapter 54th of Merivale.

In this chapter, he examines into the reasons of,
what certainly does seem very extraordinary, the
uttter tameness and unresisting patience with which
the Romans submitted to the tyranny of Nero.

This tyranny as he observes, was supported by
no military force of importance, for all the legions
were quartered on the frontiers, or in the remote
provinces, and merely a few cohorts were under the
hands of the Prince.

Neither was he upheld by any superstitious
theory of divine right, or hereditary sanctity. " The
" Emperor of the Romans stood absolutely alone at
"the head of his people. He had no society of
"tyrants of his own class, like the slave-owner, to
"support him. He had no foreign allies, like an
"autocrat in modern Europe, to maintain his
"authority as a bulwark to their own."

Yet the noblest and best men of Rome submitted

1865. it would seem, without a thought of resistance, and
performed suicide at his command just as if they
had been Japanese. Some of the reasons which
Merivale assigns for this submissiveness appear
unsatisfactory. The really essential part of his
explanation of the phenomenon may be summed up
thus — "First, the corrupt morality of the age
"perverting all ranks and classes, was, above all, the
"the cause of that patient endurance of tyranny,
"which so lamentably distinguished it. With the
"loss of self-respect, engendered by merely selfish
"indulgence; men lose that keen sense of wrong,
"even when inflicted on themselves, which nerves
"the hand of resistance more vigorously than fear
"or pain."

There was no sympathy between the aristocracy
and the multitude, who hardened and brutalized by
the sports they delighted in, felt no indignation at
cruelties which did not *touch* them.

The best men who were the principal victims,
were disposed by their favourite philosophy, rather
to a quiet scorn of tyranny than an active resistance
to it.

This chapter contains also an interesting account
of the philosophy in vogue at Rome at that time,
and of its principal cultivators ; and a very pleasing
notice of the poets of Nero's day, Persius and
Lucan. Also some account of the Jews at Rome ;
and an inquiry into the curious questions connected
with the persecution of the " Christians " under
Nero. I confess it never struck me before, but it is
certainly strange, that only three years after St.

Paul's arrival at Rome, the Christians there should 1865. be described by Tacitus as " Ingens multitudo," and as generally hated by the people; and yet that they should so soon have been forgotten, that Pliny in the reign of Trajan, should appear to consider their religion as something new and unheard of. Merivale remarks also, that the martyrologies of the Church do not specially commemorate any Christian as having suffered in the persecution.

March 25th.

Having received the two last vols. of Carlyle's "Frederick," read in the 5th vol. his account of the battle of Prague. Carlyle has certainly a remarkable talent for describing localities; above all, for giving us clear and vivid ideas of the most commonplace, tame and featureless country, his narratives of battles are very good, and he makes us understand the scenes of action better than most writers.

Monday, 27th March.

We began our preparations for going to London, and for the painting of the library.

Tuesday, March 28th.

Heard of dear Mrs. Mills' illness; sent a servant to Stutton to inquire.

Thursday, March 30th.

Read in Carlyle's "Frederick the Great," the

1865. account of the battle of Leuthen.　I omitted to note
that on the 29th I read the battle of Rossbach :
both are admirable in his peculiar way.

Read part of 8th satire of Juvenal.

Friday, March 31st.

A visit from Lady Cullum and the Rickards'.
The Abrahams dined with us and we had a very
pleasant evening.

Saturday, April 1st.

The Abrahams went away after Breakfast.
Fanny went with Mrs. Abraham to Mildenhall and
returned alone.

Strolled about the grounds a good deal :—
examined Ferns, finished chapter 57th of Merivale :
the civil war between the partisans of Vespasian
and Vitellius and the death of the latter.　It is
curious to observe the contrast between Otho and
Vitellius, if ever there was aman who threw away
his life foolishly and recklessly in mere wanton
impatience, it was Otho : whereas his successor
could not muster courage enough to dispatch
himself in circumstances, when, if ever, suicide
might have been justifiable.　It was seldom that a
Roman noble failed in that kind of courage.

Sunday, April 2nd.

Read St. John, chapter 4, in Greek, with Alford's
commentary

We went to London by Ipswich: arrived at 48, Eaton Place, at 7.30. p.m.

Mild and wet, stayed at home all day. Mary and Katharine came to see us.

Read part of chapter 60th of Merivale, to the death of Vespasian. There seems to have been much good in this emperor, and he was certainly very useful; Merivale is decidedly partial to him, and shows satisfactorily that the parsimony, for which he was so much ridiculed and reviled, was urgently required on public grounds.

Walked out with Fanny, called on Sarah Hervey at Lady Rodney's, then on Susan MacMurdo. Visits from Charles and Mary after they had been at Court. Read the remainder of chapter 6th of Merivale, containing the short reign of Titus. The first eruption of Vesuvius, and the death of Pliny, are very well told.

Distressing news of Scott's illness. I wrote to him. Went out in the carriage with Fanny. Went to the Athenæum and walked back. Dear Cissy arrived. We dined with the Henry Lyells:—a pleasant party.

1865.   Saturday, April 8th.

A better account of Scott. A visit from Mrs.
Adair.

_____

Sunday, April 9th.

Kate MacMurdo dined with us : her Father and
Mother and Edward came in the evening.

Read in Herman Merivale's "Historical Studies,"
the article on "The Landscape of Ancient Italy as
shown in the Pompeian Paintings,"—very good: but
I differ from him on some of the botanical questions
involved. In particular, I believe the Chesnut to
be an aboriginal native of Italy ; and I do not see
sufficient reason for believing that the vegetation of
Italy had, within historic times as he supposes, a
more northern character.

In the general reflections, in opposition to the
"Positivist" notions with which Herman Merivale
concludes this chapter, I must heartily concur.

_____

Monday, 10th April.

Splendid weather. Visited Katharine. Walked
to the Zoological Gardens, saw the Penguin, &c.
Norah and Margaret Bruce dined with us, and the
Adairs came in the evening. Dear Cissy went
away.

Finished chapter 62nd of Merivale's History,
concluding the reign of Domitian. The "Flavius
ultimus" was quite as cruel and detestable as Nero,
but less contemptible, as not abandoned to
sensuality. I cannot understand Merivale's evident

wish, to find some defence or excuse for this savage, 1865.
—although he ends with acknowledging that the
attempt is hopeless.

The 21st anniversary of our engagement—a day
of happy memory.

We drove round Hyde Park and visited the
National Gallery. Visits from the Henry Lyells,
the Miss Boileaus, and the MacMurdos.

We called on the Benthams,—pleasant. Went
to the South Kensington Museum and saw the
pictures.

A dinner party at home—Sir Frederick and Lady
Grey, Sir Edmund and Lady Head, the Mac
Murdos and the Charles Lyells.

Read some more of the "Life of Sir Joshua
Reynolds" and the remainder of chapter 63 of
Merivale's "Roman Empire," containing the greater
part of the reign of Trajan. It is a great pity that
we have not more personal details and anecdotes of
this excellent emperor, who really seems to have
deserved all his reputation. It is a provoking
caprice of fortune, that while we know so much
more than is worth knowing of Caligula, Nero, and

D

1865. Domitian, we seem to have little information about Trajan, except in an official way.

Miss Phillips came to luncheon. Mary Lyell and Miss Ticknor called later ; saw John Moore and others at the Athenæum.

Wrote to Scott. Drove with Fanny round Hyde Park and admired the flowers ; read the description of the shield of Achilles (part of the 18th book), in Lord Derby's "Homer,"—very spirited.

We called upon the MacMurdos. I had an interesting talk with *him* on his plan of defence. Afterwards I went to the Museum of Practical Geology and the Athenæum. We dined at the Youngs.

Read the Article on Vambery's "Travels in Central Asia :" an entertaining account of a most adventurous and extraordinary journey, performed by a very extraordinary man.

Went to see Leech's sketches, on view at Christie's.

We dined with the Charles Lyells. Met Sir Roderick Murchison, Mr. Donne, John Moore, and others. Mr. Donne said he had met Mr. Vambery 1865. (the Central Asiatic Traveller) at Lord Houghton's, when he was in England. He (Vambery) said, one of the greatest difficulties he had in keeping up his disguise all the time he was in those countries—in keeping up the character of a Turkish Dervish or holy man—was that of maintaining the appearance of perfect apathy and composure under all circumstances. Besides the danger which was peculiar to himself, and owing to his disguise, he was exposed in common with his Musulman fellow-travellers to many and various changes; and to have any chance of passing for a holy man, it was absolutely necessary that he should appear utterly apathetic;—avoiding all appearance, not only of fear, but of the excitement which danger usually produces in any European.

Murchison spoke of the Wahabees, those most fanatical of all Musulmen, who now have entire possession of the central parts of Arabia; and of Mr. Palgrave's extraordinary travels among them. He says, "the one deadly, unpardonable sin in the estimation of the Wahabees, is smoking tobacco. Murder, violation, are trifles; but if you are caught smoking, there is no hope of mercy for you;—off with your head!" Certainly, as I remarked, they could not have found the prohibition of tobacco in the Koran, but it seems these ultra-orthodox Musulmen take the liberty of adding to the articles of their religion. Moore suggested that this is a liberty not quite unknown in European churches.

1865.   Friday, April 21st.

Visit from Mary and Miss Ticknor. Finished
chapter 64 of Merivale's "Roman History." It
contains an interesting review of the men of letters
of the 'Flavian period."—(That is from Vespasian
to Trajan, both included), with observations on
the state of society as inferred from their writings.

———

Monday, April 24th.

Brilliant weather with a sharpish wind. Lady
Bell came to luncheon. Visited the Boileaus,—
very pleasant.

———

Wednesday, 26th April.

Long talk on business with Mr. Weight. Shocking news from America. President Lincoln murdered.

Bentham's evening party at the Linnean Society
at Burlington House. The rooms beautifully decorated with abundance of fine flowering plants
in pots from Kew, especially Azaleas and Rhododendrons. Mr. Wallace's splendid collection
exhibited, of stuffed bird's skins from New Guinea
and the Indian Archipelago : most beautiful and
curious ; a great many curious species of new or
rare plants, dried, and of fruits and other vegetable
products, brought from South Tropical Africa by
Dr. Welwitsch : fine dried species of that splendid
and anomalous Anonacea Monodora Myristica, in
flower, and separate fruits of it ; a new and distinct

species of Monodora :—also the first genuine species 1865.
of Nutmeg, Myristica, that has been found in Africa.
A large specimen of that strangest of all plants, the
Welwitschia, the stem looking rather like an exceedingly ill-shaped and distorted saddle, than like
any vegetable production.

Thursday, 27th April.

Wrote letters to Jeffrys and Scott about Cockfield.
We drove out to Norwood and established ourselves
at the Queen's Hotel, Upper Norwood, not very far
from the Crystal Palace.

Norwood, Friday 28th.

We spent most part of the day in the Crystal
Palace, enjoying the works of art in the Greek,
Italian, and Renaissance Courts. and the beauty of
the Tropical plants in the conservatory. The
beauty of the large Ferns, arborescent and otherwise
remarkable, growing in pots or tubs a little over the
surface of the water of the great tank in the
Tropical department. The Tree Ferns—Dicksonia
Antarctica and Squarrosa; Cyathea dealbata,
a Cibotium, and perhaps some others—not yet of
great height, but in splendid health and beauty
of development, with their glorious fronds, having
ample space as well as plenty of moisture. The
stems of the Dicksonias and Cyathea already
clothed with a thick shaggy coating of air roots.
Asplenium lucidum also in great luxuriance and
beauty, in the same situation, with its large bright
green and very glossy fronds. A fine Marattia,

1865. probably cicutæfolia. Bananas of beautiful and
vigorous growth in the same conservatory ; also
several fine Palms, especially a grand plant of Sabal
Palmetto.

Norwood, Saturday, April 29th.

Brilliant weather, but very cold east wind. We
visited Norwood Cemetery, and saw the tombs of
my Uncle and Aunt (Sir William and Lady Napier) ;
we returned home. Dear Minnie (Mrs. John Napier)
dined with us.

Monday, 1st May.

Drove out with Fanny and Minnie. We called on
Lady Smith. We had a dinner party at home :
Lombes, Henry Lyells, Bowyers, Minnie, Sarah,
John Herbert, Edward ; all very pleasant.

Tuesday, May 2nd.

Read the remainder of chapter 65th of Merivale ;
this last part of it contains a sketch of the last
struggles of Jewish nationality against the Roman
power, under Hadrian.

We drove out to Saville House at Twickenham,
dined with the Richard Napiers, and spent the
night there. Mrs. Napier well. Poor Richard
Napier very much depressed.

Wednesday, May 3rd.

From Twickenham we drove to Hampton Court,

and called on Mrs. Ellice. She and Miss Ellice 1865.
went through the galleries with us. We returned to
Twickenham to luncheon and to London to dinner.

_____

Thursday, May 4th.

Miss Julia Moore came to luncheon. I drove
out with her and Fanny and saw Lear's pictures.
Met Mrs. Abraham there.

Went to the evening meeting of the Linnean
Society. Bentham read or rather spoke a paper on
" Several New Genera and Species of Leguminosæ
from Tropical Africa." Some of these are remarkable, particularly one (Baikiæa, I think is the name)
which has the largest flowers yet known in the
family of Leguminosæ, each individual flower being
nine inches long ; it is a gigantic woody climber,
ascending to the tops of the trees in the high
forests. Another is a Dalhousiea, identical (if I
understood rightly) with a species found in the
Himalaya. Bentham said that the great woody
climbers of tropical forests, such as the Bauhineæs,
hardly ever flower in cultivation.

There has been for many years a Bauhinia in the
great Palm-house at Kew, which has grown luxuriantly, and spread extensively, but never flowered.

_____

Friday, May 5th.

Drove out with Fanny to Mr. Eddis, to whom I
sat for two hours for a drawing ; then we went to
Mary's.

Finished reading " The Life and Times of Sir

1865. Joshua Reynolds," by Leslie and Taylor. It is very entertaining, very pleasantly written, and gives a most copious supply of information, in a chatty way, about Sir Joshua and his contemporaries, and almost all his contemporaries who were of any sort of note—or notoriety—both men and women, were included in the lists of his sitters; and of all of them Mr. Taylor gives us some account. It is true that most of this anecdotic information is not new to one who is well up in " Boswell " and " Horace Walpole," but it is pleasantly brought together and well arranged.

Sunday, 7th May.

Finished chapter 66th of Merivale's " Romans under the Empire," comprising the reign of Hadrian. It is a very good chapter. Hadrian was one of the most interesting of the Emperors, and it is a great pity that we have not fuller information about him. Though not so good a man in his private character as Trajan, he was certainly superior in talents and knowledge, and his private vices do not seem to have been allowed to interfere with his public administration. Merivale "is dis- " posed to regard the reign of Hadrian as the best " of the imperial series. His defects and vices were " those of his time, and he was indeed altogether· " the fullest representative of his time, the complete " and crowning product as far as we can judge, of " the crowning age of Roman civilization."

Merivale's account of this Emperor's visit to Athens and Alexandria, the two great seats of learning and universities of the Roman world in 1865. those days, is very interesting. But he hardly touches on the remarkable splendour of this reign in reference to the fine arts, at least to sculpture.

———

Monday, May 8th.

We went to a small evening party at the Edward Romilly's.

———

May 9th, Tuesday.

Met Boxall at the Athenæum and went with him to Colnaghi's to see our damaged picture of Lady Blake.

———

Thursday, 10th May.

We went to Brighton.

———

Friday, 11th May.

Dined with Lady Louisa Kerr. Met there Sally (Mrs. Hanmer Bunbury) and her daughters.

———

13th May.

We returned to London. Dear Minnie and Sarah spent the evening with us.

———

Sunday, 14th May.

We went to the Chapel Royal at Whitehall, and heard Charles Merivale preach, instead of Kingsley whom we expected.

———

Monday, 15th May.

Read chapter 1 of Lecky's "History of Rationalism."

1865.   Our dinner party consisted of the Bishop of
London and Mrs. Tait, the Kingsleys, Benthams,
Charles Lyells, Minnie Napier, Sir John and Miss
Boileau, MacMurdos and Edward and many others
in the evening.

Friday, 19th May.

Mr. Power and Mr. Byrne dined with us.

Saturday, 20th May.

I attended the Levee.   Met Lord Stradbroke and
others.   We dined with the Miss Moores.   Met
John Moore, Edward, Charles and Mary, Mr.
Boxall, Mr. Spedding.   Mr. and Mrs. Brookfield.

Finished reading chapter 1 Lecky's "History of
"the Rise and Influence of the spirit of Rationalism
"in Europe."

This chapter contains a history of the superstitions relating to magic and witchcraft, and more
particularly of the cruel persecutions arising out of
those beliefs: written in a pleasant style and treated
in a very interesting and instructive manner. The
leading ideas, which this chapter is intended to
establish seem to be these,—that the belief in
witchcraft is a natural development of the belief in
malignant beings of super-human power and
activity: that in all ages and countries, where the
belief in devils or evil spirits and in the *terrific*
aspects of religion has prevailed with great force
and intensity, an intense belief in magic or witchcraft has followed as a natural consequence, and
has led to frightful cruelties, and that the decline

and ultimate disappearance (except among the 1865. vulgar) of this belief, was produced by no one great writer, by no great discovery, by no great controversy, but by the gradual and almost insensible progress of the spirit of research and free inquiry, or (as the author calls it) of rationalism. All this is very skilfully worked out, and in a very interesting manner.

Sunday, 21st May.

Finished Merivale's "History of the Romans under the Empire," an excellent and very instructive book. The last, the 68th chapter includes the reign of that most admirable of princes, Marcus Aurelius, and concludes with some observations on the state of the Roman Empire at the time of his death, and the symptoms of decay and decline which were already, even then beginning to be apparent in various departments of it. The view which Merivale gives us of the times of Marcus Aurelius, and even of his personal career, is a melancholy one ; he represents his reign as a constant struggle against constantly increasing dangers—against a mass of evils which was becoming irresistible, and which he felt to be so. Although he (Merivale) is far from unwilling to do injustice to the good emperor, he has not drawn so pleasing and so satisfactory a picture as Matthew Arnold in his beautiful essay on Marcus Aurelius.

Of Merivale's work generally, I have said enough in my notes on the several chapters as I have read them since last November when I began the third

1865. volume. It is a very satisfactory history. In many
places, indeed, I have thought that he shows too
strong Cæsarian tendencies, and too much a
prejudice against the Aristocratic senatorian party;
but on the whole I do not think he is unfair. The
chapters on the literature, manners and social state
of different periods are remarkably good. Merivale
also appears to me very creditably free from clerical
prejudices ; not one of those who would acknowledge no virtue in a Pagan.

Monday, May 22nd.

We called on Lady Cullum. I sat to Mr. Eddis,
and afterwards went to a tea-party at Mary's.
Edward dined with us.

Tuesday, May 23rd.

Very hot morning. Thunder-storm in the afternoon. Went to the Horticultural gardens. Splendid
display of Rhododendrons. Then to the South
Kensington Museum.

Thursday 25th.

We dined with Sir John Boileau : — a very
pleasant party. Lord William and Miss Compton,
Sir Thomas and Lady Truebridge, Mr. and Mrs.
Shirley, etc.

Friday, 26th May.

We went with Sarah to Miss Coutt's Fête at
Highgate :—beautiful grounds. We dined with
Minnie Napier.

Lady Bell and Miss Richardson came to luncheon. We drove with Lady Bell to the Horticultural Gardens, then to Mr. Eddis's, and I sat to him for the completion of my portrait.

Monday, 29th May.

Still the same beautiful weather. Katharine came to luncheon with us. Called on Miss Johnstone. A dinner party at home. Mr. and Lady Anne Lloyd, the Lister Kayes, Douglas Galtons, MacMurdos, Minnie, Sarah, Sir George Young, Mr. Esmeade and Edward.

Tuesday, 30th May.

Twenty-first anniversary of our happy marriage. Thanks be to God. Mr. Sam Smith came to breakfast with us; drove with Fanny. Mary's evening party.

Wednesday, May 31st.

Professor Schimper of Strasburg, the great bryologist breakfasted with us. His conversation extremely good. He is about to make a tour through Scotland, Wales and Ireland, with a view both to the study of Mosses and the observation of glacial geology.

He is engaged on a grand comparative work on "The genera of Fossil Plants," similar in design, seemingly, to what I once began.

1865.   We went to Buttery's to see our picture of Lady
Blake.   We went with Susan MacMurdo and Kate
to Norah Bruce's party.

Thursday, June 1st.

We dined with the Hutchings's : a very pleasant
party,—Mr. Tighe and Lady Louisa, Mr., Mrs.
and Miss Preston, Admiral Yelverton.   Went
afterwards to a pleasant little party at Minnie's.

Saturday, June 3rd.

Drove out with Fanny.   We went to the private
view of the British Institution.   Met the Boileaus.
Edward dined with us.   I read to-day an interesting memoir of Lord Elgin, (unpublished I believe)
by Dean Stanley.

Read also part of Wallace's very important and
beautifully illustrated paper "On the Phenomena
of Variation and Geographical Distribution as
illustrated by the Papilionidæ of the Malayan
Region," in the last part of the Linnean transactions.   Spent some time in writing out fairly the
notes which I had made (chiefly from Weddell's
Chloris Andina) the day before at the Linnean
Society.

Monday, 5th June.

A most beautiful day.   Wrote on business to Mr.
William Paine and to Scott.

We drove to Combe Hurst and spent some hours
very agreeably in that lovely spot, now in its full
beauty.   Mr. Smith and his daughters very agreeable.

Thursday, 8th June.

Went to Harley Street to see Charles and Mary just setting out for Germany. Drove with Fanny down to Fulham to see the MacMurdos. Minnie dined with us.

Friday, 9th June.

Dined at the Adairs.

Saturday, 10th June.

Went with Fanny and Katharine and Mrs. Byrne to the flower show at the Horticultural Gardens.

We dined with the Boileaus. A very pleasant party. Sir John and Miss Kennaway, Sir John Hippisley, &c.

Monday, 12th June.

Read a good deal of Cowper's "Letters," and some of Trevelyan's "Cawnpore."

We dined with Minnie. A pleasant party. Countess Teleki, Lady Raleigh, &c.

Tuesday, 13th June.

Called on Mr. Babbage.

The MacMurdos, Kinlochs, Katharine, Mrs. Byrne and Edward came to afternoon tea with us.

Wednesday, 14th June.

Went to see the Zoological Gardens. Called on Mrs. Byng. Katharine and Harry spent part of the evening with us.

Thursday, 15th June.

Drove to Kew Gardens, and spent two hours and a half very pleasantly there. They are delightful in

1865. this sultry weather; beautiful, shady trees, deliciously green lawns and slopes. Two houses occupied by the tropical Ferns, an exceedingly rich and instructive collection, and many of them fine specimens.

Victoria tankhouse ; that beautiful and curious plant, the Nelumbium speciosum, in flower, and bearing at the same time its singular fruit in various stages of growth.

Most beautiful blue and crimson Water-lilies from India.

Great Palmhouse. Many Palms of several different genera, of superb growth ; Bamboos, very large; Pandanus, several species, with immense leafy crowns.

Great Greenhouse (the new one), very fine specimens of Australian and New Zealand plants.

Araucarias (four or five species) Dammaras, Dacrydiums, Banksias, Aralia crassifolia, &c., many of them of great size ; also fine Dicksonias, and Cyatheas. Some beautiful Himalayan Rhododendrons, with very large white flowers, delightfully sweet scented.

———

Friday, 16th June.

Making preparations for our journey. Katharine and dear little Rosamond came to see us. Edward dined with us.

———

Saturday, 17th June.

We left London for Holyhead by the 7 a.m. train from Euston Station : reached Chester at noon, and

Holyhead soon after two: had a very fine passage. 1865.
William Napier met us at Kingston, and Emily at
Dublin.

_____

Staying at Palmerston near Dublin with the
William Napiers: glorious weather, we drove with
William and Emily to Lucan.

_____

Palmerston.

We went into Dublin with Emily, and spent
nearly five hours at the great Exhibition. Had
luncheon there. Admired the sculpture and picture
galleries, especially some of the foreign pictures.

_____

Palmerston.

Very hot and bright. Spent the morning in quiet
reading.

_____

Expedition with Emily to Castletown and
Celbridge.

They had a dinner party. Lady Campbell and
her daughter, Mrs. Smith. Lady Campbell very
agreeable.

_____

Palmerston.

William went away to the camp at the Curragh.
We went again to the Dublin Exhibition. Met
Lady Campbell there. Saw with her the gallery of
pictures by old masters. Many fine things. Also
the Water-colour gallery.

E

Friday, 23rd June.
Palmerston.

Heavy showers. Fanny and I drove to Glasnevin
and saw the Botanic Garden. Very pretty grounds,
beautiful flowers, and a fine collection of rare
plants in the hothouses.

———

                                    Saturday, 24th June.
Palmerston.

We went to the Curragh of Kildare to see
William. We went by railway from Clondalkin to
Kildare, where he met us. Saw the round tower
and ruins of the Abbey. From thence by car to
the Camp on the Curragh. We had luncheon
with William, who returned with us to Palmerston.

———

                                    Monday, 26th June.

We left our dear friends at Palmerston, and
drove to Salthill Hotel at Monkstown, near Kings-
town harbour, where we spent the evening.

———

                                    Tuesday, 27th June.

We left Ireland; had a very fine passage to
Holyhead. Went on by the mail train to Chester.
After luncheon drove about the city in an open
carriage. Saw the Cathedral and two other
Churches; the tablet in the Cathedral to John
Napier. Queen's (Railway) Hotel at Chester
remarkably good. We left Chester by the mail
train at 2 p.m., and reached London at 6.25.

———

                                    Thursday, 29th June.

A wet day. Dear Susan came to breakfast with
us, and Minnie and Sarah came afterwards.

Very wet. Count Von Randwyck breakfasted with us. Visit from Mr. Matthews.

———

[While we were at Chester, I made a day's excursion to see the different places connected with the Bunbury family. Sir Charles having a cold did not accompany me. I first went to Stanney, a property which had come into the family of the Bunburys in the reign of Edward III., the head of the family, David de Bunbury, having married the heiress of Stanney. It was the principal estate of the Bunbury family till the middle of the eighteenth century, when Sir William Bunbury inherited Mildenhall and Barton. It was sold by Sir Henry Bunbury to the Ecclesiastical Commissioners in 1859 for £108,000. The Vicar, Mr. West who had been given the incumbency by Sir Henry, received me with great cordiality, and gave me a pane of a window associated with the family, an interesting relic. He also lent me an account of the parish, which he had written out; this I had copied. Mr. West told me the people of the place were much attached to the Bunbury family, and wished much that Sir Charles would buy the place and again become master of it. I also went to Thornton-le-Moors, a place also, I believe, belonging to the Bunburys, as there are various tablets in the Church to the memory of various members of the family].

FRANCES J. BUNBURY.

———

Satusday, July 1st.

We went to Saville House, and dined with the
Richard Napiers, who are in a so-so state of health.
Catty Napier staying with them.

———

Sunday, July 2nd.

Very fine and hot.   We went to morning Church.
Minnie and Count Von Randwyck came to luncheon
and Susan afterwards.   Edward dined with us.

———

Tuesday, July 4th.

Extremely hot.   Shopped with Fanny.   We
ended with the South Kensington Museum, where
we saw the cartoons and the collection of min-
iatures.   Farewell visit from Katharine.   Minnie
and Edward dined with us.

———

Wednesday, July 5th.

We travelled post down to Barton, by Woodford
Epping, Bishop Stortford, Chesterford, Newmarket,
stopping half-an-hour at Rivers's nursery garden at
Sawbridgeworth.   Arrived safe at home—all well,
thank God.

———

Friday, July 7th.

Much rain with warm weather between.   Walked
round the Park and made various arrangements.
Susan, Mrs. Byrne and Mrs. Power arrived.

Being at home again, I returned to Lecky's
" History of Rationalism," and read (for the
second time) part of chapter 3rd, on the Fine Arts
in relation to religion and rationalism.

A long talk with Cooper on Mildenhall business. Walked over my farm. Read some more of Lecky, finished that part of his 3rd chapter which relates to the fine arts. The chapter is entitled "On the Æsthetic, Scientific and Moral Development of "Rationalism"; and it consists indeed of those parts which do not indeed appear to be very closely connected together. But this first part, in which he traces the history, progress and modifications of art (especially of painting) as affected by the prevailing states of religious belief in different times, is remarkably interesting.

He touches, with admirable truth and beauty of sentiment and expression on the beneficial effects of the veneration—perhaps one may say adoration—of the Virgin Mary; I have . extracted this passage in another of my note books. Then, taking up the history of painting, he shows how in the middle ages the art was exclusively religious, both beauty and truth of representation being subordinate to the religious conventional notions; how, in the time of the highest perfection of painting, the religious and the æsthetic elements were harmoniously blended and co-operated to the production of the greatest works, and how latterly the religious sentiment disappeared, and the love of beauty became the ruling idea of art.

Read the 14th Satire of Juvenal.

———

Heard of the death of poor old Mrs. Scott.

1865.   Finished the 14th Satire of Juvenal. The earlier
part of this to about verse 140, where he treats
of the mischief done to the morals of the young
by the evil example of their parents, is admirable ; in a high and noble strain of morality,
powerful, true, and unexaggerated.   There are
some lines in it which well deserve to be remembered as *texts*.   After this, when he runs on for the
remaining 200 lines on one single branch of his
subject—" The Lust of Wealth,"—he overloads his
subject and becomes tedious.

Wednesday, 12th July.

Fanny, Susan and I came over in the open
carriage to Mildenhall, leaving Mrs. Power and Mrs.
Byrne at Barton.

Thursday, 13th July.

Mildenhall.
Wet and chilly weather.   Heard of Lord Alfred
Hervey's defeat at Bury.   Read part of Lord
Fortescue's book on Public Schools for the Middle
Classes.

Friday, 14th July.

Mildenhall.
A beautiful day.   Fanny and I attended the
funeral of old Mrs. Scott. Visited the Boys' School.
We all drove to the hill, and had a pleasant
stroll there.

Saturday, 15th July.

A splendid day, very hot.   Fanny and I went

through the workhouse with Mr. Lovelock, and 1865.
looked at the Church.

Mr. Lott came to luncheon with us. We returned
here to Barton.

————

Monday, 17th July.

A splendid day and very hot. Went on with
Lecky on " Rationalism," and finished the part of
it which relates to (what he calls) " The Scientific
Development of Rationalism," that is the progress
of physical science as affecting, and affected by the
progress of " Rationalism," or freedom of enquiry in
religion and morals. He shows the extraordinary
impediments thrown in the way of all scientific
enquiry, and all advance in knowledge by the
religious notions and the Church system of the
middle ages ; giving some amusing examples of the
sort of theories on scientific subjects, which grew
out of the teaching of the Church.

He says very justly, that " it is indeed marvellous
" that science should ever have revived amid the
" fearful obstacles that theologians cast in her way."

He then shows how the rise and progress of
scientific enquiry, especially in the 17th century and
since, has gradually acted upon and modified the
general habits of thought even upon subjects not
belonging properly to the domain of physical science.
In particular how every natural phenomenon that
was uncommon, exceptional or startling, was at
once as a matter of course, referred to a special
interference of Divine power ; how for example, it
was long believed that comets, eclipses, earthquakes,

1865. and even storms of unusual violence, were quasi-miracles, not the regular normal results of the natural laws ordained by God for the working of the universe; and how as one after another of these phenomena has been brought within the domain of natural laws, the tendency to see a miracle in every unexpected event has been gradually diminished.

Lecky also makes some excellent remarks on the contrary tendency, too common now-a-days, to mistake natural *laws* for the true *causes* of things, or to put them in the place of final causes. For instance—the motions of the heavenly bodies are perfectly explained by the laws of gravitation; but of the *cause* of gravitation, we are perfectly ignorant.

I read also part of Satire 5 of Juvenal.

———

Tuesday, 18th July.

A long talk of business with Scott.

Finished Satire 5 of Juvenal. It is not a particularly interesting Satire, but contains some curious bits of incidental information, concerning dishes, &c., and some curious hard bits for the commentators to exercise their ingenuity upon; and it discloses a strange state of manners. It is hard to believe that in the age of the highest civilization of Rome, and in the highest class of society, there should have existed such brutal coarseness of manners as to allow of such insolent distinctions as described; that a "great man" should invite his clients to his table for the very purpose of insulting and annoying them.

Read some more of Lecky. Read in the evening 1865. some of Lord Campbell's "Life of Lord Hard-wicke."

Thursday, 20th July.

Finished chapter 3rd of Lecky's "History of Rationalism." It is a disproportionately long chapter, and I do not understand why the author has thought fit to throw it into one, as it treats of three different subjects, as distinct, seemingly, as most of those to which he has allotted separate chapters. The latter portion treats of religious terrorism*—of the modifications which have taken place in the belief of future punishment. He shows how great a place was occupied in the Christianity of the earlier and middle ages by the doctrine of a bodily existence after death, and of actual material and eternal torments; how intensely this belief was "realized;" how incessantly it was dwelt upon by the clergy, and especially by the monks of the middle ages; how they dwelt, and urged their hearers to dwell, upon the contemplation of it in every possible form; and what effect it had on the minds and characters of men. I believe he is quite right in thinking that the cold, merciless cruelty which has so often been remarked as characteristic of monks and priests, must have been much fostered by the habit of dwelling on such contemplations. But I am afraid there may be some doubt whether "religious terrorism" is so obsolete, or the *odium theologicum* so nearly extinct as he supposes.

* By the way, I think Lecky is rather too fond of words in "*ism*."

1865.   Friday, 21st July.

Read the 11th Satire of Juvenal. Expecting a friend to dine with him, he here gives him some account of the entertainment he is to expect, and takes the opportunity to declaim against the luxury and extravagance of the time, and to contrast it with the rude simplicity of the early Republic. A hackneyed topic of rhetorical complaint in all civilised ages and countries. There are some pretty touches;—in particular that of the rustic boy :—

" Sûspirat longo non visam
tempore matrem.
Et casulam, et notos tristis
desiderat hædos."

LETTER.

Barton,
July 21st, 1865.

My Dear Katharine,

I have little doubt you and your children are enjoying the repose and freshness of the country in Scotland as I am here. I have been much interested in all I have heard of your proceedings, and this letter, let me tell you, is a good deal intended to get you to write direct to me. What a lovely photograph of dear Rosamond that is which you sent to Fanny. I have heard of Frank's initiation as a sportsman. I used to follow my Father as a spectator of his sport when much younger than he is.

I can remember when my great ambition was to be old enough to be allowed to carry a gun ; but *when* that time came, my feelings and wishes had 1865. entirely changed; so much the worse for me, perhaps.

It happened curiously, that the rains began the very night after we returned home; the country must have been excessively parched before, but the abundant rains and frequent thunder storms since then have done a wonderful deal of good. In particular, they have saved the "root crops,"—a fact of which, as you do not inhabit the eastern counties, you perhaps do not appreciate the importance. Our lawn and park are now as beautifully green as in spring, the foliage is very rich, and many plants are putting out a second crop of flowers; the Wisteria, in particular, has a very handsome second crop. The Catalpa tree in the arboretum is blossoming most beautifully just now; quite one mass of flowers; and it is visited by such innumerable bees that when one stands by it on a sunny morning, the noise is as if one were in the midst of a number of bee-hives. I get out for half-an-hour before breakfast most mornings, and it is very delightful when the sun is shining bright and the flowers and grass all sparkling with dew. I have seen the Humming Bird, Sphinx Macroglossa stellatarum, three times in the last ten days, and one time I saw it settle for a moment on a wall, which is a sight I never saw before. Our Ferns, most of them, are doing very well, and I have bought three new ones, Pteris (or Pellacæ, geraniifolia), Aspidium trifoliatum and Dicksonia antarctica. But the pretty little Hymenophyllum, which

1865. we brought from Wales last year, is dead, and Lycopodium Selago in a doubtful state.

(July 22nd.) This very damp weather is, I am afraid, bad for poor Mrs. Power, who seems to suffer very much at times, but she is wonderfully cheerful. Dear Susan looks very well ; it is a great pleasure to have her here. Fanny is, as usual, extremely busy, and at present especially deeply immersed in deliberations concerning the colour of the library curtains, a weighty question which is not yet settled. The painting of the library has been perfectly successful, I think. Lady Blake has made her appearance in her renovated state, looking as fresh as the morning and very handsome. She and Lady Sarah are not yet hoisted into their permanent places, but stand looking stately and gracious in the middle of the library. It will be some time yet before the room is habitable. The two mirrors are up in the entrance hall, and certainly improve its appearance.

Fanny sends her love, and pray give mine to your husband and children, and kind remembrances to the Miss Lyells ; and believe me

Ever your very affectionate Brother,

CHARLES J. F. BUNBURY.

JOURNAL.

Sunday July 23rd.

A very warm moist relaxing day. We went to morning church.

Finished reading vol. 1 of Lecky on "Rationalism." The last chapter of this volume takes up 1865. in effect the subject of the last part of the preceding chapter; it treats of the "antecedents" (or pre-disposing causes) of persecution, and in particular of the doctrine of "exclusive salvation." This doctrine — that endless torment is the necessary destiny of all who are outside of a particular church, is indeed the true logical preliminary to persecution. Lecky treats this subject very ably, and shows how earnestly this intolerant doctrine was held, not only by the Early Fathers and by the Church of the middle ages, but also (with a very few exceptions) by the early Reformers. As connected with this he gives a very curious but very painful history of the belief in the inevitable damnation of unbaptized infants.

Monday, July 24th.

Visits from Arthur Hervey, Lady Cullum and Mrs. Adair. Mr. and Mrs. Abraham and Mr. Lott came to stay with us. Mr. and Mrs. Sutton and Patrick Blake dined with us.

Tuesday, July 25th.

Most beautiful weather. I walked round the garden with Mrs. Abraham. Visit from Shafto Adair. Studied and arranged some of the Malvaceæ of my Herbarium.

Mrs. Abraham and Susan sung charmingly in the evening.

1865.   Wednesday, July 26th

Splendid weather.   The Abrahams and Mr. Lott
went away.   We dined with the Arthur Herveys.
A pleasant party.   The Dean of Canterbury and
Mrs. Alford, the Abrahams, &c.

Friday, 28th July.

Walked over my farm.   Lady Cullum dined with
us.

Tuesday, 1st August.

Very cold.   Mrs. Power and Mrs. Byrne went
away.   We drove to Stowlangtoft, and saw Mrs.
Rickards, Mr. Rickards being unfortunately not
well.

I read the rest of Juvenal's first Satire.   I have
often thought that in many points of manner and
social condition, Rome in the best period of the
Empire had a striking resemblance to England in
the present century ; and in Juvenal himself, I very
often meet with sentiments and remarks very
applicable to our own times.   But there are some
things in the manners of Imperial Rome, which
strike us as very strange and foreign to all our
notions.   For instance, the *Sportula*, that daily
dole of food and money distributed at the doors of
the rich—distributed, not to what we should call
the poor, but to the dependents and humble friends
of the great family ;   and even if we may believe
Juvenal, to impoverished nobles.   In this Satire
he represents men of old family, and even men
who had held dignified offices, crowding for their

share of the dole, while the poorer clients were 1865.
mainly dependent on it for their ordinary sub-
sistence.

———

Very wet and cold.

Went to the Petty Sessions at Bury. Finished
the first chapter of Vol. 2nd of Lecky's " History of
Rationalism." This chapter—on " Persecution," is
excellent. He traces the shocking history of the
practice of persecution for opinions, from its rise in
the early Christian Church, almost as soon as the
Church acquired the power of persecuting, through
the times of its most horrid excesses, to its
decline and virtual extinction in modern times. He
shows with great force, how closely and logically it
is connected with the doctrine of exclusive salvation.
He points out that the great leaders of the Reformation, with very few exceptions are as intolerant in
theory as any of the Papists; and that nevertheless
the Reformation had a most powerful though
indirect effect in promoting toleration.

" To the Reformation is chiefly due, the appear-
" ance of that rationalistic spirit, which at last
" destroyed persecution."

But is it quite true, as Lecky says, that Queen
Elizabeth persecuted Papists on purely religious
grounds, independently of politics? Lecky does
not enter at all into the history of persecution before
the establishment of the Christian State Church
under Constantine; and therefore does not attempt
to clear up what always seems to me a most curious

1865. question—the causes of the specially and exceptionally ferocious persecution of the Christians by Pagan Rome.

_____

Thursday, 3rd August.

The Maurices and the Abrahams came to stay with us. The Arthur Herveys dined with us.

_____

Friday, 4th August.

Weather improving. Mr. and Mrs. Bentham arrived here. I walked round the arboretum and garden with Mr. Bentham. Lady Cullum dined with us. The Abrahams went away in the evening.

_____

Saturday, 5th August.

Walked through the garden and arboretum with the Benthams. We went—a party of six to Hardwick, and walked about it. Mr. and Mrs. Image dined with us.

_____

Sunday, 6th August.

Went to morning Church. Showed some of my collections to Mr. and Mrs. Bentham. These two days I had the advantage of much conversation with Mr. Bentham. He is a man of much and various information, and has travelled extensively in almost every country of Europe.

_____

Monday, 7th August.

Mr. and Mrs. Bentham went away. They are very pleasant people.

Read Gifford's "Notes on the First Satire of 1865. Juvenal." Walked to my farm. Pleasant talk with the Maurices and Susan.

Tuesday, 8th August.

Went with Susan and the Maurices to Ickworth, and saw the Arthur Herveys.

Wednesday, 9th August.

Clement arrived in the evening.

Thursday, 10th August.

Walked to my farm—wheat well stacked. Saw reaping machine at work on Paine's land. Susan and Mr. Maurice went to Mildenhall and returned to dinner.

Friday, 11th August.

The Maurices went away after luncheon. For a week past I have had the advantage of conversation with Mr. Maurice who has been staying with us: and I am sorry that I have not much that I can distinctly record. But in truth, the impression which his conversation leaves on my mind is rather a general impression of pervading wisdom and goodness, than anything specially and distinctly to be remembered. His talk is not fluent nor strikingly brilliant, but one feels somehow the better for it.

Dear Cissy and the children arrived: the latter very flourishing and Emmy very tall.

Monday, 14th August.

A terribly wet and stormy morning, very fine

F

1865. afterwards. The Bishop of London and Mrs. Tait arrived, having come by the Sudbury line. We walked about the grounds with them; afterwards arrived Mr. Tyrell and the Abrahams. The Arthur Herveys dined with us.

Tuesday, 15th August.

Wretched wet weather, but we went out after luncheon. Fanny went with Mrs. Tait in the pony carriage. The Bishop, Mr. Abraham, Mr. Tyrell and I, walking. We visited my farm, the church, &c. Lady Cullum dined with us, and the Douglas Galtons and Lady Louisa Kerr arrived.

Wednesday, 16th August.

We went, a party of eight, to Ickworth, and had luncheon with the Arthur Herveys, saw the big house, and then went with Arthur Hervey to Bury, and saw under his guidance the two churches and the Abbey ruins.

Thursday, 17th August.

Wretched wet weather. The Bishop and Mrs. Tait and Mr. Tyrell went away. I heard of the death of Sir William Hooker; our home being constantly full of company, my reading has been sadly desultory and unsettled lately: besides that the damp steam-bath weather has relaxed and unnerved me, but I have now managed to finish chapter 5 of Lecky on the "Secularisation of Politics." In this he traces the progressive effect of "Rationalism" in the theory and practice of Government (that was not the mere reign of brute force) was under the 1865. influence of the clergy, and directed entirely by ecclesiastical considerations down to the time of the French Revolution, and indeed to our own days. A remarkable part of this chapter is, where he shows the anti-monarchical, and even democratic aspect which the teaching of the Church of Rome occasionally (though perhaps exceptionally) assumed: as, in particular, in the writings of Mariana and some other Jesuits. In contrast to this, he traces forcibly the history of the Anglican Church, and shows how uniformly through its most brilliant ages, it was the enemy of liberty and the advocate of slavish submission.

Friday, 18th August.

The Douglas Galtons went away. Walked to the gravel pits. Sally and her daughters arrived.

Tuesday, 22nd August.

Dearest Susan went away to our great sorrow. Mr. and Mrs. Phelps arrived.

Wednesday, 23rd August.

The children's party.

Thursday, 24th August.

Wretched wet weather. We received the sad news of the death of my dear old friend Mr. Rickards. Lady Louisa Kerr and the Phelps went away, also Herbert and little Henry.

Saturday, 26th August.

A beautiful day.   Henry arrived.

———

Sunday, 27th August.

I finished the reading of Lecky's "History of the
Rise and Influence of the spirit of Rationalism in
Europe." In the last chapter on "The Industrial
History of Rationalism," the most interesting
portions are those in which he traces the changes
which have taken place since the early Christian
times in opinion and legislation relative to slavery :
to interest on Money, to the Jews, and to theatrical
entertainments and actors. It seems perhaps a
little droll to class an account of the stage under the
head of Industrial History : at least, though actors
do no doubt in reality work very hard, we are not
used to look on their occupation as one of the forms
of Industry ; but Lecky's sketch of this branch of
History is curious and interesting. His sketch of
the persecutions endured by the Jews is still more
interesting and even touching. The latter part of
the chapter, where he gets on the subject of political
economy as affecting general politics, is much less
to my taste. But he winds up the chapter and the
work with a remarkable avowal. After enumerating
with great force and justice the many and great
benefits to mankind which the advance of "Rationalism" has secured : the destruction of the belief
in witchcraft and of religious persecution, "the
"decay of those ghastly notions concerning future
"punishments, which for centuries diseased the
"imaginations and embittered the character of

"mankind : the abolition of the belief in the guilt of 1865.
"error which paralyzed the intellectual : and of
"asceticism which paralyzed the material progress
"of mankind"; after all this, he fairly admits that
there is a shadow resting on this otherwise brilliant
picture. The decline of individual force and
greatness : the decline of strong convictions and
the enthusiasm of self-sacrifice : these form the
shadow; and Lecky concludes with the avowal,
that "it is impossible to deny that we have lost
something in our progress."

On the whole, this book of Lecky's appears to me
a good and a pleasing one, as well as a very
interesting one. It has a considerable affinity to
Buckle's unfinished work; one sees at once that it
belongs to the same school of thought : but it
appears to me a better book, and decidedly more
pleasing: shewing much less arrogance, much more of
the spirit of reverence in the author. Yet it seems to
have made much less noise than Buckle's book,
much less *sensation:* perhaps for the very reason
that it is less arrogantly dogmatical, that it does
not attack received opinions with such audacity.
The main fault I find in Lecky's book is, that he
does not explain in the beginning what he means by
Rationalism. What he does mean evidently is, the
habit of inquiring and searching and judging by
those methods of investigation with which our
reason supplies us : instead of believing implicitly
and without inquiry whatever we are told.

Sally and her daughters went away. I went into Bury, attended a meeting about the Cattle Plague, and the Petty Sessions.

Visit of Arthur Hervey, who had come for the funeral of Mr. Rickards.

LETTER.

Barton,
September 4th, 1865.

My Dear Lyell,

I was very glad indeed to receive your letter, and much interested by it; besides being very glad to hear of your safe arrival in England. We had heard of your proceedings from time to time through Mary's letters as well as some capital ones from Leonard, sent us by Katharine; but I was still much in the dark as to the geological results of your expedition, and am particularly glad to have your own account of them.

You seem to have made a very interesting tour. The earth pyramids near Botzen, I remember very well—that is as one remembers things at a distance of 38 (no, 37) years; and though when I saw them I had scarcely even heard of moraines, yet I can quite believe, from what I remember of the locality, that such may have been their origin. As for the Glen Roy terraces, which you illustrate, by the facts you observed, on the Aletsch glacier, the whole theory is to me very puzzling; but I will look back to a paper of Jamieson's in the *Q. J. G. S.*, which I

remember, seemed to give me clearer ideas on the 1865.
subject than anything else I have read.

Leonard, I have no doubt, was a very pleasant
companion, and must have enjoyed the tour
immensely ; I am sure I should at his age ; and I
dare say the impressions of it will last as long as he
lives.

I am glad you heard all about us from dear
Susan. I was very sorry indeed to part with her.
I hope we shall soon hear of her safe arrival at
Florence.

We have had a very sad loss in the death of our
dear old friend and neighbour Mr. Rickards, one of
the most charming men I have ever known. We
had grown so intimate during the last five years,
had had such constant and confidential intercourse,
and we always found him such a true and cordial
friend, so kindly and cheerful, so full of various
information and taking so much interest in everything that was interesting to us, that his
death makes a blank that will not easily be filled
up.

We have had some pleasant society, as you will
have heard from Susan. I was really surprised to
find how agreeable Bentham is : not only a mine of
botanical knowledge, but so full of various information and anecdote and imparting it so pleasantly.
It was very unlucky that Fanny was so ill during
the two days they were with us, that she hardly saw
anything of them.

I am very anxious to know who is to succeed Sir
William Hooker at Kew. There can be no sort of

1865. doubt that Joseph Hooker ought to have it, but I am very much afraid it will be jobbed.

I hope you will enjoy your visit to Birmingham, where I suppose you will enlighten the meeting with the results of your Swiss researches. Much love to dear Mary. I am ever,

Your very affectionate friend,

CHARLES J. F. BUNBURY.

P. S.—Fanny is gone over to Mildenhall for the day with Cissy and the children.

JOURNAL.

Monday, 4th September.

A splendid day. Very hot. Henry went to London and returned late. Fanny, Cissy and the children went to Mildenhall and returned to dinner.

Tuesday, 5th September.

Another most beautiful day. Received a letter from Mr. Boehm about my Grandfather's picture, and wrote to him in reply.

We had a dinner party:—the James Blakes, Admiral Blake, Lady Cullum and the Suttons.

Wednesday, 6th September.

We went to Woolverstone (Mr. Berners') with our own horses part of the way, and posting the rest—arrived late, met there Sir John Boileau and his two daughters.

Thursday, 7th September.

Glorious weather. In the morning had a very pleasant ramble through the beautiful Fernery on the cliff, with Mr. Berners and Sir John Boileau. In the afternoon drove with Fanny and Miss Boileau to Stutton, and saw the Mills. Lord Stradbroke arrived.

Friday, 8th September.

I went through the gardens with Mrs. Berners.

In the afternoon we went—a large party—to the Ipswich race course—saw the conclusion of the rifle contest, also the distribution of the prizes, also heard the speeches. In the evening Fanny went with Mrs. Berners to the ball at Ipswich.

Saturday, 9th September.

We returned home, and found dear Lady Napier there.

The Kingsleys arrived: as charming as ever. I walked with Kingsley through the arboretum, garden, &c.

Sunday, 10th September.

Excessively hot. We all went to morning Church. Pleasant talk with the Kingsleys. Kingsley read a sermon to us in the evening.

Monday, 11th September.

A splendid day—very hot. Spent a very pleasant morning with Kingsley, looking over dried plants.

We took the Kingsleys into Bury in the afternoon, and went through the Botanic Garden, &c.

1865. with them. The Wilsons and Suttons dined with us.

_____

Tuesday, 12th September.

Another extremely hot day. Again a pleasant morning botanizing with Kingsley. Mr. and Mrs. Bowyer and young Anstruther arrived.

The Arthur Herveys dined with us. Mrs. Bowyer sang charmingly.

_____

Wednesday, 13th September.

Splendid weather. Another pleasant botanical morning with Kingsley, though he was not well. We had a pleasant drive with Mrs. Kingsley and Mrs. Bowyer. Lady Cullum and the Maitland Wilsons dined with us.

_____

Thursday, 14th September.

Same glorious weather. Dear Henry went away. So did young Anstruther. Morning spent as before, with Kingsley.

_____

Friday, 15th September.

Weather as beautiful as ever. Dear Cissy and her darling children went away to our great sorrow. Morning spent as before, with Kingsley. Much pleasant talk with Mr. Bowyer and Kingsley.

_____

Saturday, 16th September.

I went with Mr. and Mrs. Kingsley and Mr. Bowyer to Ickworth. Weather hotter than ever. Mrs. Bowyer sang in the evening.

_____

We went to morning Church. Kingsley read us a sermon in the evening. I had a stroll and a pleasant talk with Kingsley before going to bed.

————

Our dear friend Mr. Kingsley went away in the morning, as also the Bowyers. Dear Lady Napier went after luncheon, Mrs. Kingsley and Rose remaining.

Read some more of "Jules César."

In the last nine days I have enjoyed much delightful conversation with Charles Kingsley. I should have enjoyed his society as much as ever, had I not been very uneasy about his health. He is indeed evidently in a very uncomfortable state of health, though I trust not in an incurable one. But his mind is clear and vigorous, and his conversation if not quite as animated, as rich and various and instructive as ever. His interest in botany and natural history generally, continues to be eager. I have noted down in another book some of his remarks on these subjects. He is very favourably disposed towards Darwin's speculations, without plunging into them with the headlong zeal of Huxley and Lubbock. In reference to art, I observe in him still more strongly (I think than before) a distaste to Gothic architecture, and the art of the middle ages, connected with a general dislike to the mediæval institutions and modes of thought. This of course is connected with the dislike (in which I share) to the mediæval tendencies of the modern

1865. High-Church party and their efforts to revive the superstition as well as the art of those times.

While we were amidst the ruins of the Abbey at Bury he said that he was glad to see the ruins of the buildings which had been raised through cheating and plundering. He would not allow that the Monastic orders had been a protection to the poor against the tyranny of mere force, nor even that the Church in the middle ages had been the best form of Christianity which the then state of mankind admitted. He rather seemed inclined to hold (with Michelet) that the Church at those times had conspired with the nobles against the poor. Perhaps he may even yet have a tendency to be too extreme and unqualified in his conclusions,—though much less than in his "Alton Locke" days. He has a bad opinion of the Welsh, as a people, and not much better of the Irish. The Cornish, he says, are in their physical characteristics a quite distinct and peculiar people, and he does not believe that they are identical with the Welsh. In the course of an interesting talk about Cornish superstitions, he said he believed that the superstition about *spectral* or *dæmoniacal hounds* (which are called Wisht Hounds in Cornwall and Devonshire) originated in the strange sounds made by wild fowl in winter nights. The cry of flights of wild fowl, and especially of wild swans, he says, is strangely like the cry of hounds, and in a wild wintry night might easily suggest the idea of a spectral chase.

Then, in each country, this primary idea becomes mixed up with the idea of the punishment of some personage who is a special object of popular hatred.

Kingsley admires Moore's poetry :—spoke en thusiastically of the merit of his versification—the music of his poetry, praised particularly the exquisite rhythm of my favourite—"Silent, oh Moyle, be the roar of thy waters."

He contended, that authors in general are, in their real characters and everyday life, very different from what we should infer from their writings. I brought forward Moore as an instance to the contrary. "But then," said Kingsley, "consider Moore was an Irishman, and therefore comes under no rule."

He holds that *mixed* races of mankind are the best : that all races are improved by mixture. He highly admires William Pitt (the second) as I do too. He agreed with me that the generation of Pitt and Fox and their contemporaries and the next generation also, the men who grappled with the French Revolution and with Napoleon, were eminently great :—seemed generations of giants.

Kingsley said that the Prince and Princess of Wales, in their house at Sandringham, are most kind and pleasant hosts, very attentive to the comfort of their guests, and very simple and natural in their manners. The Princess charming. The Prince thoroughly amiable.

Kingsley thinks that Thackeray, in his lectures, has been too hard upon the first three Georges : I confess I doubt.

Kingsley said : I wish I could believe in the regeneration of a nation,—but I see no example of it in history.

LETTER.

Barton,
September 18th, '65.

My Dear Katharine,

1865.   I was very glad to hear of the re-union of
the scattered members of the family in London
from Scotland and the Alps, and that Leonard
returned to you safe and sound, and no doubt much
delighted with his tour, which (after they left
Kissingen) must have been a very interesting one.
I hope you spent your time agreeably in Scotland.
I was very sorry to part with dear Susan; I
enjoyed her society very much, and wish she could
have stayed on with us till now, as I am sure she
would have enjoyed being here with the Kingsleys
and the Bowyers. We have indeed spent the last
twelve days delightfully: the first three of them at
that beautiful place of the Berners's, Woolverstone,
where we met Sir John Boileau and those two
charming girls, his daughters, Mary and Theresa;
when we returned from thence we found dear Lady
Napier (Sir George's widow) here with Cissy; and
the Kingsleys arrived immediately after. I have
found them as charming as ever, and the only drawback to the pleasure of their society has been that
the state of Kingsley's health makes one very
uneasy. I fear indeed that his health is very bad;
I only hope and trust that it may rally before long:
I feel very anxious about him. But his conversation is still delightful, and his mind active on a

variety of subjects. I had a great deal of pleasant 1865.
talk with him on botany and natural history in
general (besides other matters), and we spent part
of almost every morning in looking over my
collections. He went away this morning, but Mrs.
Kingsley and Rose will remain with us a few days
longer. Mr. Bowyer I think you must have met
here or at Mildenhall, as he has been often with us;
he is a very agreeable man, of great ability and
high cultivation; and Mrs. Bowyer is a pleasant
person, and sings charmingly.

I was very sorry to part with dear Henry and
Cissy and their darling children: and the dear
little things were so sorry to leave us (especially to
leave "Aunt Fanny") it was quite touching.

We have had more than a fortnight of the most
splendid weather imaginable, cloudless skies and
extraordinary heat: just the weather for the
arboretum. I have pitied the dwellers in London
during this weather; but perhaps you will think I
am like the sailor, pitying "those unhappy folks
ashore" in a gale of wind.

It is a great pity that Susan left us before this
fine weather began; I am afraid she will always be
incredulous about our sunshine. It is odd that the
heavy rains with us, began exactly as the harvest
began, and ended as it ended. The Fungi which
overran the whole place during the wet weather, so
that we really could hardly walk, either on the lawn
or in the woods without treading on them, have
all disappeared.

This hot dry weather suits me exactly: I have

1865. seldom felt in better health, and Fanny also seems to be well,

The quantity of insects in our garden is indescribable,—especially those buzzing flies *(Syrphi)* which look so like bees, also nettle-tortoiseshell butterflies, and in the evening midges and gnats ; but, strange to say, *no* wasps. That pretty creature, the humming-bird hawk-moth (Macroglossa stellatarum), has been more plentiful here this year than I ever saw it before, and Rose Kingsley tells me the same has been the case at Eversley.

I have read very little lately I am sorry to say, but since I finished Lecky's very interesting book, I have begun the Emperor's "Histoire de Jules César," which as far as I have yet read I find very dry and heavy. I hope I shall read more as the days shorten.

We are now expecting the arrival of my grandfather's portrait, as the bargain is concluded, and I have heard from Mr. Boehm that he is ready to send it. I hope he has not let me have it for less than its value.*

Pray give my love to Harry and your dear children.

Believe me ever your very affectionate Brother,
CHARLES J. F. BUNBURY.

* He gave £100 for it.

JOURNAL.

Fanny, Rose, and I went to Bury to the luncheon 1865.
given to the Queen of the Sandwich Islands.

Thursday, 21st September.

The Barton rent audit. We went by railway to
Ely. Arthur Hervey and his daughters Sarah and
Kate joining us at Saxham; dear Mrs. Kingsley and
charming Rose left us at Cambridge to our great
regret. We and the Herveys went to the *Lamb* at
Ely.

Friday, 22nd September.

We and the Herveys went by railroad from Ely to
Peterborough, thenin an open carriage to Crowland ;
saw the fine remains of the Abbey and the curious
bridge. Returned to Peterborough to luncheon and
back to Ely by train :—a very pleasant excursion.

Saturday, 23rd September.

Ely.
We attended morning service in the Cathedral,
and afterwards went through it with Dr. Goodwin*
the Dean. Returned by 12.50 train to Bury.
Parted with the Herveys at Saxham.

Monday, 25th September.

Dear Clement went away and we were alone.

* Present Bishop of Carlisle.

1865.   Had a talk with Scott on business. We walked (Fanny and I) about the grounds together, and gave several directions about trees. As we are now again alone, I resumed the catalogue of my herbarium in which I have got as far (on Endlicher's system) as the Orchidaceæ.

Wednesday, 27th September.

We drove to Ashfield and saw Lady Blake. Then called at Stowlangtoft and saw Mrs. Rickards, the first time I have seen her since her great misfortune.*

Friday, 29th September,

We dined at Hardwick. Met the Arthur Herveys.

Saturday, 30th September.

Drove to Ixworth, and saw the Church.

Sunday, October 1st.

A beautiful day. We went to morning Church. We took a walk together in the park.

Monday, October 2nd.

Scott brought me a very good account of the Mildenhall rent audit. Read John Stuart Mill's Essay on "Civilization" in his "Dissertations and Discussions;" a very powerful exposition of the evils accompanying our modern European — (and especially English) civilization ; namely:—the growing

* Her husband's death.

insignificance of individuals relatively to the mass; 1865. the decay of individual energies: the decline of the heroic virtues: the weakening of the influence of superior minds over the multitude : the continually increasing power of charlatanerie. I go along with him very heartily in much that he says on these topics, but when he comes to the remedies, he is not so satisfactory. With his furious invective against the system of our Universities, I have no sympathy at all; but the sketch with which he concludes of the idea of a University Education, is worthy of attention.

Tuesday, October 3rd.

Fanny read me Arthur Hervey's sermon on the death of Mr. Rickards :—excellent. Read the principal articles in the new number for (October) of the *National History Review*. The most interesting of them is that on the Gare-fowl or Great Auk, tracing very carefully the curious history of the gradual extinction of that remarkable species.

LETTER.

Barton,
October 5th, 1865.

My Dear Susan,

Very many thanks for your letter from Viareggio, which has given me very great pleasure. I am truly glad that you have had so much satisfaction in reading Lecky's book, and I have

1865. been very much interested by your remarks on it.
I quite agree with you in your general high
appreciation of the book, and especially in what
you say of "his reverential spirit, modesty and
candour." Indeed, it is a book that gives me a remarkably agreeable impression of the author, as well
as a very high estimate of his powers. Few, I think
could have written on the subjects which he has
treated, with so much fairness and candour, so little
of a sectarian spirit, and so much readiness to see
the good points even of those whom he is obliged
most strongly to condemn. It is uncommonly free
from that "superstition in avoiding superstition" (as
Bacon expresses it) which one often sees so strongly
exemplified in those who call themselves great
" Liberals."

Lecky's work evidently belongs to the same
school of thought as Buckle's, but it appears to me
a better book, as well as a much more pleasing one,
yet it seems to have made much less sensation ;
perhaps for the very reason that it is less arrogantly
dogmatical, and does not attack received opinions
with so much audacity. I am charmed like you
with what he says of the effects of the homage paid
to the Virgin Mary ; this, I copied into my extract
book ; and I was much struck also by his remarks
(especially at p. 184, vol I.) on the decay of dog-
matic and the increase of practical Christianity.

I found that Kingsley had not yet read Lecky's
book, which I strongly recommended to him, but
he had come to the same conclusion as to the origin
of religious persecution, that it springs naturally

and logically out of the intense belief in exclusive 1865. salvation. Still I cannot think that it is in general so free from selfish motives as this would imply. I think there is much of pride and self-will in religious persecution. A persecutes B not simply because he thinks his opinions erroneous and wicked, but because B is stiff in opposing his favourite opinions.

Your remarks on Greek and Gothic architecture are very interesting, and I doubt not, very true. I quite agree with you that it is a book which would have delighted your dear Father, and which might perhaps have led him advantageously to re-consider some of his opinions. I hope Fanny will some of these days find time to read it. I am now reading the Emperor's book, " The Histoire de Jules César." I do not admire it. The first book, which is more than half of the first volume, is a mere dry abridgement of the Roman History down to Cæsar's time ; very dry ; unrelieved by great general views, by brilliant sketches of individual character, or even by brilliant paradoxes ; very different from Montesquieu or even from Michelet.

The second book, which includes the youth of Cæsar, is more interesting, but it is a panegyric, not a history ; so extravagantly partial as to destroy all feelings of confidence. This period has been much better treated by Merivale, who is also Cæsarian enough in all conscience. The only striking things in the Emperor's book are those passages in which he reveals the *purpose* of the book : the purpose of showing that certain men such as Cæsar (and by

1865. implication, the two Napoleons) have been *provi-
dential men*—men of destiny—whom it was impious
to resist. A very convenient doctrine for *them*.

So much for books. As soon as I have finished
" The Emperor," I hope to read over again Montesquieu's "Grandeur et Décadence des Romains"
and then to begin Gibbon.

I need not attempt to tell you any news, as I know
Fanny sends you a Journal. I shall only say that I
heartily wish you could have been with us during
last month, which was one of the pleasantest I can
remember to have passed. It was delightful having
the Kingsleys with us so long and in such a quiet
way ; and dear Lady George Napier too, who is so
charming ; and the Bowyers ; besides Henry and
Cissy and their children, who left us in the middle
of the month. And after all these were gone, our
three day's trip with the Arthur Herveys to Ely,
Peterborough and Crowland (of which I do not doubt
that Fanny has sent you a full account) was extremely agreeable. I like the Kingsleys, all three of
them—more than ever ; I feel a real affection for
them. The only drawback to the enjoyment of
this visit was, that Kingsley was in a very uncomfortable state of health, and I felt very uneasy
about him in that respect ; I trust, from the last
accounts, that he is now much better. We have
now been nearly a fortnight alone, and the time has
not hung heavy on hand. But somehow or other
I have done much less in this quiet time than I
expected to have done, and ought to have done.

On the 9th we expect dear Minnie and Sarah,

and Kate MacMurdo; and on the 14th Charles and Mary, whose visit, though short, will be a great treat; and the week following, we shall be full of gaiety.

It is really a pity you left us just before the fine weather set in, so that as far as the skies were concerned, you had a very unfavourable specimen of Suffolk. All through September, almost without interruption, we had the most glorious weather possible; cloudless skies, and great (sometimes intense) heat, day after day. It was exceedingly enjoyable here, under the shade of our trees and in airy rooms; and it has agreed wonderfully well with me; ever since the beginning of September, I have felt myself in better health than for the whole year before. Yet it has certainly not been a healthy season generally.

We are now wishing for rain, not only because the pastures are burning up, and the turnips cannot grow, but because there is a vague hope that wet and cold may stop this terrible cattle plague. Happily this parish is *as yet* exempt from it, but we live in daily apprehension.

Give my love to dear Joanna and say I hope to write to her soon.

I am ever your very affectionate brother.

CHARLES J. F. BUNBURY.

October 9th.

JOURNAL.

Monday, October 9th.

Read Trench's two Lectures on "Gustavus

1865. Adolphus," and on the "Social Aspect of the Thirty Years War;" both good, the second very striking and impressive. It is an awful picture of the effects of the terrible war upon Germany; not only the immediate misery and desolation, but the long-lasting check to prosperity and civilization. It is very well done.

Went on with my catalogue of plants. Dear Minnie and Sarah and Kate MacMurdo arrived.

Thursday, October 12th.

Read chapters 6th, 7th and 8th of Montesquieu. The 6th is that masterly chapter in which he, Montesquieu, expounds with such remarkable clearness and force, the system of iniquitous aggression —of treachery and violence—by which the Romans conquered successively all the nations with which they were acquainted. It was a marvellously skilful and well-organised system of wickedness; and yet so cloaked under the forms of justice, that it is quite possible that those who carried it out may not have been sensible of its immorality.

Heavy rain all the morning. Went on with the catalogue of the Herbarium. Took a walk with Fanny and Minnie. Read " Palamon and Arcite " to the ladies in the evening.

Friday, 13th October.

Read chapters 9, 10, and 11 of Montesquieu, his judgment of Cæsar and Pompey appears to me more true than Napoleon's; and indeed than Arnold's, who is disposed to exaggerate the other way.

Saturday, 14th October.

Dear Charles and Mary Lyell arrived—very late
—looking very well.

Sunday, 15th October.

Went to church with Minnie and the girls. Had
a pleasant walk with Charles and Mary.

Monday, 16th October.

Took a walk with Charles Lyell. Sir John and
Miss Mary Boileau and Admiral and Mrs. Eden
arrived, also Mr. Lott ; the Arthur Herveys dined
with us.

Tuesday, 17th October.

Most of our party went to lunch at Ickworth. I
had a pleasant walk with Charles Lyell. We had a
large dinner party. Mrs. Abraham and the girls
sang charmingly. Charles Lyell is engaged on a
new edition of his " Principles," and is at present
deeply engaged with the examination and discussion
of a new hypothesis, (by a Mr. Croll) concerning the
submergence of the Northern hemisphere, and the
production of great changes of temperature through
the variation of the earth's orbit. He seems to
think that there is something in this hypothesis ;
but he says the problem how to account for the
great geological changes of temperature is so extensive and complicated and involves such a variety
of elements as to be almost beyond solution by
human faculties.

He had been much interested and gratified by

1865. the geological examination he made during his late
tour, of the famous "Earth-pillars" near Botzen,
and of the glacial lake on the Aletsch glacier, (this
as illustrating the parallel roads of Glen Roy), but he
does not mean to bring out any separate papers on
them, which I am sorry for.

Wednesday, 18th October.

Charles and Mary went away.   Had a pleasant
walk about the grounds with Sir John Boileau.
Sir James Simpson and the Wilsons dined with us.

Thursday, 19th October.

Excessively wet and stormy.   The Boileaus and
the Edens went away.   Read chapters 15-17 of
"Montesquieu," excellent remarks on the cruelties
of the Emperors, and of the tendency of the
Romans generally to cruelty, on the reasons which
made the populace of Rome favourable to the worst
Emperors, on the progressive corruption and decay
of Roman Institutions, and therewith of Roman
strength, and on the mischief done by the inno-
vations of Constantine in particular.   Read in the
new number of the *Edinburgh Review*, the articles
on the Peleponesus, by Edward.

Saturday, 21st October.

A very wet morning.   Major Tyrrel, Charles
Berners and Robert Anstruther went away.   The
Thornhills came to luncheon.

LETTER.

My Dear Katharine,

Many thanks for your kind letter. I have 1865.
felt for you during this anxious time of the illness of
your dear boys, for though it is certainly a very good
thing to get the measles over in childhood or earliest
youth and have done with them, I know that there
must be much need for care and ground for a
certain degree of anxiety while they are going on.
I hope Leonard will soon completely recover his
health and strength, and that if Arthur and
Rosamond catch the disorder, they will pass
through it well and speedily. I am very sorry to
hear that you have got a severe cold in addition to
your other troubles.

We have had a very gay week and a very pleasant
one. Charles and Mary's visit, though much too
short, was a great treat: and so was that of Sir
John Boileau and his daughter. Our dance on
Thursday was exceedingly successful, and seems to
have given pleasure to all concerned, young and
old. Sarah and Kate MacMurdo and Miss Kinloch
and Sarah Hervey looked and danced charmingly
and seemed very happy: and there were several
other nice girls and agreeable women and a gentleman-like set of men. The dancing was kept up
with infinite spirit. I did not go to the Bury ball,
but Fanny did, and stayed very late, and I am
happy to say she does not seem the worse for all
this dissipation, though of course she was very

1865. sleepy the next evening. Now we are still a party
of eight, for besides dear Minnie and Sarah and
Kate, Mr. Kinloch stays with us till tomorrow, and
Miss Kinloch and Sir George Young till Saturday.

Lord Palmerston's death is certainly a serious
event. I should think he will be much regretted by
those who knew him personally. As a statesman,
I know he was one of your special aversions. I
cannot agree either with your estimate of him or with
that of his worshippers. To my thinking he was
not only a singularly successful politician, but a very
sagacious, very prudent and very useful one, but
scarcely a great statesman : at least, not of the
class of Somers and the two Pitts and Sir Robert
Peel. He may perhaps be classed with Godolphin
and with Walpole. Remember, however, that
Buckle lays it down as a maxim, that "not truth,
but expediency should be the object of a statesman."
Lord Palmerston seems to have anticipated the
maxim of Buckle.

Earl Russell, I see, is to be the new Prime
Minister: very properly, for his great experience and
long and steady devotion to the Whig cause, give
him a good claim ; but I can hardly think that he
will be able to hold the Ministry together very long.
Gladstone, I take it, is inevitable sooner or later :
and I am sorry for it ; for, much as I admire his
genius and high cultivation, I dislike his democratic
tendencies and fear his crotchets. In the meantime, the next session of Parliament is likely to be
much more stormy than it would have been if Lord
Palmerston had lived.

I have lately made a considerable addition to my 1865.
botanical library; the whole of the *Botanical
Magazine*, the three series (by Curtis, Sims and Sir
W. Hooker) complete from 1787 to 1864: a most
useful book. I bought it at the sale of the books
of a poor old clergyman in the neighbourhood. I
am now going regularly through the volumes,
making notes.

(October 23rd). Here is another beautiful bright
day, after a spell of very wet and cold weather,
which made me think that winter was come already.

The rain has been very welcome to the farmers,
and the cold has not yet been severe enough to kill
off any of the out-door flowers, not even the
Pomegranate blossoms. The autumnal colouring is
very beautiful. We have certainly had a glorious
summer and autumn, and I have enjoyed them
thoroughly. Now I look forward to cosy winter
evenings and good reading time; but "*l'homme
propose*," &c.

Charles Lyell will have told you of our catching
the Humming-bird Sphinx, on a window in the
dining room. It has been wonderfully abundant
here this season.

With much love to Harry and your children, also
to Charles and Mary, I am ever,

Your very affectionate Brother,

CHARLES J. F. BUNBURY.

I forgot to tell you that I danced Sir Roger de
Coverley at our dance, with Mrs. Abraham, who
was charming, as she always is.

JOURNAL.

1865.   Went to Quarter Sessions at Bury—county business—only nine prisoners for trial.

Sir George Young went to Cambridge.   Read the remainder of Gibbon's chapter 1st and part of 2nd. Read also the *Edinburgh Review* on Palgrave's "Arabia : "   the book appears to be a valuable one, but I do not like the tone of the article which seems to be written by a Scotch Calvinist.

A furious gale of wind.   Sir George Young returned from Cambridge.   Lady Rayleigh, Mr. and Miss Strutt, Mr. and Lady Mary Phipps, and Mr. and Mrs. Praed arrived.   A very large dinner party.

Fanny and most of the party went over to Mildenhall to the choral meeting, and returned to dinner.   I stayed at home with Minnie and Lady Rayleigh.

A beautiful but cold day.   All our company went away except Minnie, Sarah, Kate and Miss Kinloch, Had a pleasant walk with Minnie, Sarah sang charmingly.

Read chapter 3 of "Gibbon ;" the very brilliant

and interesting picture of the state of the Roman 1865.
Empire under the "Antonines," contained in these
three chapters, is no doubt too brilliant; painted in
too favourable colours; the darker shades skilfully
omitted or softened, as in the cases of slavery,
the religious persecution and the cruelties of the
amphitheatre. The far more sombre colouring given
by Merivale to his picture of the same period, may
be more true. Indeed, the rapidity with which the
Roman Empire declined, after Marcus Aurelius,
seems to show that its prosperity, however apparently brilliant, could not have been solid.

Yet Gibbon himself (towards the end of chapter
2), acknowledges that the "public felicity" conceived
the seeds of decay and corruption; and points out
very forcibly how the universal and uniform despotism of Rome tended to produce weakness and
degeneracy. The passage at the end of chapter 3,
in which he shows how the vastness of the Roman
Empire took away from the victims of oppression
all hope of escape is justly celebrated.

Monday, 30th October.

Dear Minnie, Sarah, Katie, and Miss Kinloch
went away. I went out before breakfast to gather
nosegays for them.

LETTER.

Autumn of 1865.

My Dear Katharine,
            Fanny has not represented quite accurately
what I said about John Stuart Mill. What I say is

1865. that though I cannot go along with him in many of his opinions, I yet respect and honour him as a true philosopher, a man of enlarged and lofty intellect, of noble mind ; and I have a very different feeling towards him from that which I have towards the generality of Radicals and Revolutionists.

I have no time to write more just now, as the dressing bell has rung. Much love to your children.

Ever your very affectionate Brother,

CHARLES J. F. BUNBURY.

JOURNAL.

Friday, November 3rd.

Another beautiful day. We walked through the shrubberies to the Boys' school. Went on with the Mildenhall accounts.

Read chapter 7 of "Gibbon." The period of history embraced in these four chapters, from the accession of Commodus to the celebration of the secular games by Philip, from A.D. 180 to A.D. 248, is such a dreary tissue of crimes, follies and miseries, that even Gibbon's art and skill in narration, can hardly make it pleasant reading.

After Marcus Aurelius, the government rapidly degenerated into a mere military tyranny ; the tyranny of a fierce, licentious, lawless soldiery, domineering alike, over the unfortunate, peaceful population and over their nominal masters. If any Emperor tried honestly to do good, he was certain to be murdered by the soldiers, who always showed a decided partiality for the worst rulers. 1865. Finished the Orchidaceæ in the catalogue of my herbarium.

Visit from Lady Cullum.

———

Saturday, 4th November.

Again a beautiful day. Took a long walk with Fanny. We visited the farm, plantations, Necton, &c.

═══

LETTERS.

Barton,
November 4th, 1865.

My Dear Edward,

William Napier is coming to us (barring accidents) on the 20th of this month, and we shall be very glad—Fanny and I—if you can come to us about the same time, that is, any time in *that* week to meet him, and to help to shoot such pheasants as may be forthcoming. Do come. We hope to get the MacMurdos and some others to meet him. I suppose you have returned to Town by this time, and I hope you are the better for Brighton. We have had glorious weather these last three days, slightly frosty, just what we wanted after the heavy rains of last week, and the colouring of the trees is still very beautiful. We are alone now, Minnie and the three girls having left us on Monday. A sorrowful parting, I assure you. The three girls I should explain are Sarah, Kate MacMurdo, and Miss Kinloch ; very charming girls—all of them ; and they became very great friends. I have read with a

H

1865. great deal of interest and pleasure, your *Edinburgh Review* article on "The Peloponnesus," and have read it to Fanny, who was also much interested by it. We are now reading the *Quarterly* on "English Cathedrals," and to myself I am reading Gibbon.

Lord Palmerston must I should think be much regretted by those who knew him personally. As a public event his death is a very serious one, and I fancy the coming session will be very different from what it would have been if he had lived. Not only will there be much more determined warfare between the regular "Liberals" and "Conservatives," but I conceive there will be a tendency to disintegration in the so-called Liberal party itself. I take it that not a few of those members who were numbered as Liberals in the lists of the last elections were strictly Palmerstonians, and might not be disposed to support without reserve, a Russell-Gladstone Ministry, and still less a Gladstone-Radical one. I do not see my way through the complications, but I should doubt whether Lord Russell will be able long to hold his ground as Prime Minister. Is it known why *The Times* has taken such a decided line against him?

Ever your very affectionate Brother,

CHARLES J. F. BUNBURY.

———

Barton,
November 6th, 1865.

My Dear Leonora,

We have had much very pleasant company this summer and autumn. We have been quite alone

(for a wonder) for a week past, and we find so much 1865.
to do, that the days fly past at a tremendous pace.

I think our new Fern-house was not built when
you were last here. It is on the north side of the
wall near the greenhouse, facing the kitchen garden.
It answers very well, as far as our experience goes
hitherto, and we have several very interesting
Ferns in it, and in fine condition. The finest
are : Cheilanthes elegans, a most beautiful, delicate
plant from the mountains of Caraccas; Adiantum
formosum and trapeziforme; Onichium lucidum;
Asplenium furcatum, bulbiferum, flaccidum and vi-
viparum; Lomaria alternata; Aspidium aristatum
and caryotideum ; Nothochlaena nivea ; Polypodium appendiculatum; and the "gold and silver"
Gymnogrammes. There are also some very fine
Ferns of hardier kinds, in the old greenhouse. The
glorious summer and autumn we have had, have
been, as you may suppose, very favourable to the
garden, arboretum and so forth ; and the shade was
delightful in the hot weather. I used in that
glowing weather of September, to go every morning
before breakfast into the garden, when it looked
most lovely, when the dew was sparkling on all the
flowers, and they were visited by swarms of pretty
insects. I never saw here such an abundance of
butterflies and sphinxes as this year.

So good Sir William Hooker is gone, at a good
old age, and after doing his work most thoroughly.
Certainly there has never been in this country a
man more useful to the cause of botany, nor could
any other have been found so peculiarly fitted for

1865. the post he held at Kew.   I am exceedingly glad
that Joseph has been appointed his successor.   He
is an admirable man and has the highest claims;
but though he is a greater botanist than his
father, and a man of higher genius, I am not
sure that he is so specially qualified for that
particular post.   But it would have been a shame if
any one else had been appointed.   And now Dr.
Lindley too is gone; a valuable man too, and a
great botanist.   A death which has made a more
general sensation is Lord Palmerston's.   He was
so essentially and specially an *English* Minister,
that it is very possible he may not be so much
regretted in other countries as in this; though I
think there has been a tendency to exaggeration and
extravagance in the praises of him since his death.
I cannot look on him as a statesman of the very
highest order; but should rather be disposed to place
him somewhere about the level of Sir Robert Walpole.
There is no doubt however that his death will make
a great change in the position of parties in this
country, and that the coming session of Parliament
will be much more stormy than it would have been
if he had lived.

Give my love to dear Annie and Dora, who I
hope are quite well.   I hope they have not forgotten
Barton, and that you will bring them here next
year, and Annie will find the horses as "spiddity"
as ever.

With love to your Husband,

I am your affectionate Brother,

CHARLES J. F. BUNBURY.

Finished my examination of Mildenhall estate accounts of last quarter. We examined the Ferns. Read chapter 9 of Gibbon. This chapter on the ancient Germans and Scandinavians is full of admirable good sense and sagacity, and of Gibbon's quaint felicities of expression. He reduces to their true value the exaggerations of ancient writers, and dreams of modern ones, about the virtues of the northern barbarians and the populousness of the northern regions. Read to Fanny part of the article in the last *Quarterly* on English Cathedrals.

———

Friday, November 9th.

Read the remainder of chapter 10 of Gibbon. The period of 20 years comprised in this chapter, from the secular games to the end of Gallienus' reign, was a time, as Gibbon says, of shame and misfortune. A wretched time indeed: some provinces conquered by the Persians, others ravaged by the Goths; pretenders to the empire starting up in all directions: civil wars, pestilence and famine.

—— ——

November 10th.

Read chapter 11 of Gibbon:—the reigns of Claudius and Aurelian. The former of these seems to have been a fine character, but his reign was too short to show really what he was. Aurelian was a good soldier, but a bad ruler in peace. Under these two men the empire began to recover some of its vigour. Gibbon's narrative of the Palmyrene war is interesting: but it does not show a true sympathy with the noble Zenobia.

Wednesday, 15th November.

Went on with the Barton accounts. We drove to Stowlangtoft and saw Mrs. Rickards and visited dear Mr. Rickards' grave.

Read chapter 13 of Gibbon:—the reign of Diocletian and his colleagues. Diocletian certainly began that revolution which Constantine completed, the systematic conversion of the Roman Empire into a regular despotism on the genuine oriental pattern. He may therefore be considered to have contributed materially to the ruin of the state.

Thursday, 16th November.

Miss Richardson and Miss Joanna Richardson arrived. Wrote to Henry. Fanny had a very pleasant letter from Mrs. Kingsley, with a good account of Kingsley's health.

Friday, 17th November.

William Napier arrived. Miss Richardson sang to us.

Saturday, 18th November.

A fine day. Arranged minerals and went on with annotations on *Botanical Magazine*. Edward arrived. Lady Cullum dined with us. We gave scarlet cloaks to some of the school girls.

Sunday, 19th November.

Walked to morning church with William Napier and Miss Richardson. N. B.—The school girls in their scarlet cloaks.

Weather remarkably mild. A shooting party. I showed Miss Richardson the arboretum and the Ferns. The Abrahams and Mr. Lott arrived. Dinner party:— the Maitland Wilsons and Lady Cullum. Delightful singing in the evening.

———

Tuesday, 21st November.

Very mild weather. Looked over some of my Father's papers. Some of our party went to see Hardwick. Mr. Abraham and Mr. Lott went away.

———

Thursday, 23rd November.

A beautiful day. Fanny, Miss Richardson and Mr. Arnold went to Mildenhall and returned to dinner. Arranged and studied Mosses and some other plants. The Arthur Herveys and Abrahams dined with us. A very pleasant evening.

═══

LETTER.

Barton,
November 27th, 1865.

My Dear Mary,

I look forward with great pleasure to reading Charles Lyell's new edition of "The Principles"; though no new edition, with ever so many improvements, can ever be to me what the first edition was when I first read it. I shall never forget the impression those two volumes made on me: they seemed to open a new world of knowledge to me. That book, and Lindley's "Introduction to

1865. the Natural System," which I read about the same time, gave me the first ideas of the philosophy of natural history.

Fanny is pretty well and sends you her love : she is very busy and happy with her Library Catalogue, with which she is getting on swimmingly.

Since I finished Froude I am reading Colonel Hambley on "The Operations of War":—rather stiff reading.   There are some curious things in the last number of the *Geographical Society's Proceedings.* Baker on the "Abyssinian Tributaries of the Nile" (which is very entertaining).   Clements Markham " On the Water Supply in India, as affected by the Destruction of Forests," and two curious papers on " The Last Eruption of Santorin."

The last accounts of Mrs. Power seem rather better, and I hope she may yet be spared to her friends some time longer.

With much love to Lyell,

Believe me,
Ever your very affectionate Brother,
CHARLES J. F. BUNBURY.

JOURNAL.

Monday, 27th November.

Went with Fanny and Miss Richardsons into Bury—bought a small picture of Ladbrook's, and got the new volume of Stanley's "Jewish Church."

The Miss Richardsons (very agreeable women they are) went away early—and Edward after breakfast.

Read the story of Daphne, in Ovid's Metamorphoses.

Finished chapter 15 of Gibbon. This is one of the two celebrated chapters for which he has been so much censured ; in this he considers the secondary or human causes of the progress and triumph of Christianity in the Roman Empire. It can hardly be doubted, I think, that the various causes which he enumerates did contribute most materially to the success of the Christian Religion ; the causes he enumerates are, "exclusive zeal, the immediate " expectation of another world, the claim of miracles, " the practice of rigid virtue, and the constitution " of the primitive church,"—and also the absence of any strong belief in Paganism. I can see nothing wrong or irreligious in believing that God made use of second causes and human agency in that great work, as well as in the ordinary course of his Providence. I do not perceive anything objectionable in the *matter* of this chapter. But there is certainly something unpleasant in the tone of half-concealed sarcasm which pervades it, and especially in the affected style of mock respect which he sometimes assumes.

––––

Thursday, 30th November.

Went to London to 48, Eaton Place. Arrived a little before 3 p.m. Visits from Charles and

1865 Mary. We dined with Katharine, quietly : — a pleasant evening.

_____

Friday, December 1st.

Down to Brighton

__ ____

Brighton,
Sunday, 3rd December.

Finished 17th chapter of Gibbon : an interesting chapter, containing a brilliant description of Constantinople and a very full account of the elaborate, pompous and cumbrous system of despotism, established by Constantine; he was the real founder (at least so far as concerns Europe) of that elaborate and showy style of monarchy, with its etiquette and endless gradations of titles and offices, which has been so much in vogue in the modern kingdoms of Europe.

_____

Brighton;
Monday, 4th December.

Excessively stormy, walked on the pier, the sea rolling in grandly. We went to luncheon with Sally. Met Lady Louisa Kerr.

_____

Wednesday, 6th December.

We returned to Eaton Place. Dined at the Geological Club.

_____

Thursday, 7th December. London.

Visited Lady Grey. Met John Moore and Boxall at the Athenæum. Linnean Society Evening Meeting. Bentham and Hooker.

_____

Wretched dark weather. Wrote on business to Scott. We dined with the Charles Lyells. Met Dean Stanley and Lady Augusta, Sir Frederick Grey, Miss Moore, Mr. and Miss Grove. Mr. Maurice and the Benthams came in the evening.

Sunday, 10th December.

We went to morning service at St. Michael's. Mr. Hamilton preached on Matthew the 24th. We went to Twickenham and dined with the Richard Napiers. Found them tolerably well.

Monday, 11th December.

Went to the Museum of Practical Geology and looked at the Minerals. Mary and Katharine, the MacMurdos and Edward came to afternoon tea.

Tuesday, 12th December.

We went to Ketteringham, Sir John Boileau's.

Wednesday, 13th December.

A pleasant walk in the morning with Sir John Boileau. We went with Sir John and Theresa Boileau and saw the church. Mr. and Mrs. Digby came. We went with Sir John and Mr. Digby into Norwich, and saw the Hospital, Museum, Castle, &c. Mr. and Mrs. Evans Lombe came to dinner, also Mr. and Mrs. Irby.

Saturday, 16th December.

We left Ketteringham and came to Melton to the Evans Lombes,—most cordial and friendly.

1865.   Sunday, 17th December.

Took a walk with Mr. Lombe and afterwards took
a walk in the garden with Mrs. Lombe. We took a
drive with Mrs. Lombe and visited Lord Bayning
at Honingham:—a fine old house in a very pretty
situation—some noble Vandykes. Then visited
George Pellew and his mother at Honingham
Rectory. A dinner party at Melton.

———

Wednesday, 20th December.

During these seventeen days while we have been
away from home, first in London, then at Sir John
Boileau's, at Ketteringham, lastly at Melton with
the Evans Lombes, I have not read much:—little,
in fact, except the 18th chapter of Gibbon and two
admirable French Novels, "Le Conscrit" and
"Waterloo," both by Erckmann Chatrian.

Gibbon's 18th chapter is not particularly interesting: it relates the last years of the life of Constantine and the crimes and misfortunes of his sons.
On the 6th and 7th of December I attended
meetings of the Geological and Linnean Societies,
but did not hear anything specially interesting. At
the Linnean, however, I was very glad to see
Joseph Hooker, sufficiently recovered from the
serious illness under which he suffered since his
father's death, to appear again at a meeting of the
sort, and he did not appear to me to be looking ill.

On the 9th we dined with the Charles Lyells, and
had the pleasure of meeting Dean Stanley, who I
had much wished to see; I was very glad also to
meet the Benthams and Mr. Maurice.

LETTER.

Barton,
December 21st, 1865.

My Dear Katharine,

The Fern you have sent is what I take to 1865. be the true Aspidium aculeatum of Smith and Hooker, the most typical form of it, intermediate and almost equi-distant between the two forms or sub-species which are called *lobatum* and *angulare*. Perhaps, if anything it is rather nearer to *lobatum* The more extreme forms of the two look very different, but there are so many intermediate gradations that it is very difficult to draw any line. The *lobatum* form, I think, prevails most in Scotland ; *angulare* in the south and west of England, in Ireland, and in the more southern countries, such as Italy and Madeira. In this county we have both, but neither abundant.

I very much enjoyed the week which we spent in Norfolk, first with the Boileaus and then with the Lombes ; but I am always glad to be at home again, and we are already comfortably settled again into our old corners, and looking forward with hope and satisfaction to three weeks of (we may hope) well employed quiet. All well here, and, *as yet*, no cattle plague in the parish, I am thankful to say. At Mildenhall however, it seems to be spreading alarmingly. The weather is very dark, but very mild. The Ferns flourishing—in the houses I mean, of course. The Chimonanthus

1865. against the arboretum wall, flowering more richly than I have ever seen it since we have been living at Barton.

Give my love to dear Rosamond and her Brothers —also to Harry, and believe me ever,

Your very affectionate Brother,

CHARLES J. F. BUNBURY.

JOURNAL.

Monday, 25th December.

We went to morning Church. The servants' supper party.

December 27th, Wednesday.

A beautiful day. Wrote to Henry. Rambled about the grounds and collected some Mosses.

Finished chapter 20 of Gibbon. An interesting and very able chapter relating to the conversion of Constantine, the establishment of Christianity as the religion of the State, and the constitution of the Church in those times. There are in this chapter some very characteristic touches of Gibbon's epigrammatic manner : — "An absolute Monarch "who is rich without patrimony, may be charitable "without merit; and Constantine too easily believed "that he should purchase the favour of heaven if he "maintained the idle at the expense of the in- "dustrious, and distributed among the saints the "wealth of the Republic.". . "The liberality of "Constantine increased in a just proportion to his "faith, and to his vices."

A very stormy day.

Went on with my herbarium catalogue—grasses, —and with the annotation of the *Botanical Magazine*.

———

Sunday, 31st December.

Read *Acts*, chapter 28, with Alford's notes.

Read to Fanny, Robertson's beautiful Sermon, on the Virgin Mary.

Read part of Vigor's paper on the "Natural Families of Birds." Read to the servants one of Kingsley's Sermons.

So ends the year 1865, *Laudes Domino.*

———

Monday, 1st January, 1866.

A beautiful day. Very mild and clear. Walked 1866. with Fanny. We called at the Vicarage and saw the Admiral.* Then walked through the Dairy Grove. Went on with the catalogue of minerals. Visit from Lady Cullum.

———

Wednesday, 3rd January.

Walked with Fanny ; we chose places for young trees.

———

Friday, 5th January

Mr. Herbert, the artist, came with Mr. Gilstrap to see our pictures.

Finished reading chapter 21 of Gibbon. The History of the Arian controversy, and the struggles

* Admiral Blake.

1866. and persecutions arising out of it, afford choice food for the historian's satirical spirit. One can fancy him chuckling while he wrote of the inappreciable shades of difference between the various sects and the fierce animosities to which they give rise. Certainly the history of those times shows a lamentable want of Christian charity in all the parties concerned. Yet there was a more favourable side to all this. Gibbon makes a remarkable admission, when he says that "The caution, the delay, the "difficulty with which he (the Emperor) proceeded "in the condemnation and punishment of a popular "bishop, discovered to the world that the privileges "of the Church had already revived a sense of "order and freedom in the Roman government."

The zeal (turbulent and misdirected though it might be) of the Christians, and the influence and the independent spirit of their ecclesiastical leaders, alone broke the dead level of imperial despotism.

The history of Athanasius as sketched in this chapter is very interesting, and even as it is drawn by so unfavourable a narrator as Gibbon, one cannot help admiring the indomitable spirit and resolution of the man, little as one may sympathize with his dogmatism.

Sunday, 7th January.

Saw the first yellow Aconites. Read to Fanny Robertson's beautiful Sermon on "The Glory of the Divine Son."

Harry and Katharine Lyell and their children arrived.

Charles and Mary Lyell arrived; also Cecil, Clement and Herbert. I had a pleasant walk with Charles Lyell.

The children's party. A large gathering—very merry. I had a nice select dinner party of the Charles Lyells, Arthur Herveys and two Wilsons. Pleasant talk.

Looked over some more of the minerals with Katharine and Leonard. Much pleasant talk with Charles and Mary Lyell.

All the Lyell party went over to Mildenhall and returned to dinner.

Mr. Clark, the public orator, of Cambridge arrived, and we walked round the gardens with him. The Admiral (Blake) dined with us.

Mr. Clark, the public orator of Cambridge, Tutor of Trinity College, and Editor of the " Cambridge Shakespeare," was with us. We looked over editions of Shakespeare and other books, with him.

1866. He says that Sir Thomas Hanmer's corrections of the text of Shakespeare show a great deal of good sense and sagacity, but that they seem to have been almost entirely conjectural or arbitrary; that is— not founded on any old editions or MSS., but springing merely from his own notions of what was suitable. Sir Thomas, he says, scarcely ever quotes in his edition any old authorities; neither does he seem to have been deeply versed in the literature contemporary with Shakespeare.

Mr. Clark has seen the Duke of Devonshire's "Hamlet," the one which my Father discovered at Barton, and sold to Payne and Foss; the Duke has lent it for the purposes of the Cambridge edition; and he tells me that another copy of the same early edition has since been discovered by some one; which copy by a curious chance wanted the title page, but possessed the last leaf, which is wanting in the Barton "Hamlet." With regard to Mr. Payne Collier's famous annotated Shakespeare, Mr. Clark says he has a very strong conviction that it is a modern forgery; though not a forgery of Mr. Collier's own; perhaps, he thinks, not intended to deceive at all, but an imitation written for mere amusement by some antiquarian.

We planted some trees.

We had at dinner—Lord and Lady Augustus Hervey, Lord John Hervey, Lady Cullum and Miss Birch and the Abrahams.

Lady Augustus and Mrs. Abraham sang.

———

Wednesday, 17th January. 1866.

Dear Charles and Mary went away. Also the Abrahams and Mr. Clark.

———

Thursday, 18th January.

Captain and Lady Florence Barnardiston arrived.

———

Friday, 19th January.

Dear Katharine and her party went away.

Walked round the arboretum, farm, &c. with the Barnardistons.

The Abrahams, Lady Cullum, Miss Birch and Maitland Wilsons dined with us — delightful singers. The Barnardistons and the Maitland Wilsons went to the ball at Bury.

———

Saturday, 20th January.

The Barnardistons went away after luncheon.

———

Monday, 22nd January.

I am engaged with those difficult genera of Grasses, Panicum and its allies.

———

Tuesday, 23rd January.

Cecil and Herbert went away.

We dined with Lady Cullum—alone with her and Miss Birch, and then went to the Athenæum and heard a really beautiful and admirable lecture from Lord Arthur Hervey on "Constantine the Great." He took certainly a more favourable view of Constantine's character than I have been used to take ; he dwelt more particularly on the more favourable aspects of the reign ; but he was not unfair ; he did not conceal or gloss over the vices of

1866. his hero.   The neatness with which the results of
much study were brought within the compass of a
lecture, the arrangement of the subject, the clearness and grace of the language, delighted me ; and
still more the beauty of the moral tone pervading the
whole.   It was a worthy product of Hervey's beautiful mind.

Wednesday, 24th January.

We began to classify our library catalogue, of
which we completed the first draft on the 22nd of
December last.

Finished first book of the third part of Carlyle's
"French Revolution." The massacres of September
are told with terrible power, though with too much
tolerance.   The account of the battle of Valmy is
indistinct and probably exaggerated.

Thursday, 25th January.

We went on with the classification of our library
catalogue.

Monday, 29th January.

We marked Oaks in Sorcerer's paddock, and
walked to further end of Dairy Grove.

Arranged minerals.

We dined with the James Bevans, a pleasant
party.   We met the Arthur Herveys, Lady Cullum
and Miss Birch and the Bishop of Brechin.

Tuesday, 30th January.

Visit of the Bishop of Brechin and Mr. Bevan.
They looked at our pictures.

Thursday, 1st February. 1866.

Wet weather.

Finished Carlyle's " French Revolution." An exceedingly striking and impressive book. To my thinking it is much the most powerful of Carlyle's works, the one in which all his peculiar characteristics are developed in the most effective and advantageous manner. The style has all his peculiarities ; sometimes offensive, sometimes grotesque ; but it has the great merits of giving us a most vivid idea of the people and the times, and of stamping indelibly upon our memory the impressions which he intended.

One can hardly forget any of the scenes described in this book ; as for the moral tone of the book, it is by no means what I admire. That adoration of strength and energy, no matter how applied, and that tendency to worship success which had grown to such an offensive height in his later writings, are already apparent here, though not in the same excess. There is a stern, cold, pitiless philosophy, in his manner of speaking of the Terrorists, and their victims, which shocks one's feelings ; and the grim, sarcastic, lurking humour, which is a remarkable characteristic of the book, has now and then a painful effect. But there is no attempt to disguise or palliate the atrocities of the time. I do not know any history in which they are more impressively related.

LETTER.

Barton,
February 3rd, '66

My dear Lyell,

1866.   I thank you much for sending me Madame
Agassiz's letter to Mary, which I have read with
much curiosity and interest. The variety of new
fish and other novelties which Agassiz has discovered are not half so astonishing to me as the rapid
growth of that country. How completely Brazil
seems to be revolutionized by the one single agency
of *steam*. Madame Agassiz speaks of the voyage
from Para to the Barra de Rio Negro taking *five*
days ; when the botanist Spruce explored that
country, no longer ago than 1850, the voyage from
Para to Santorin, which is little more than half way
to the Barra, often required a *month*. Still, I should
have more confidence in observations made by men
who have been a long time stationary in chosen
spots, like Bates and Wallace and Spruce, than
in those made at steam pace.

Agassiz's observation on "glacial phenomena," in
Brazil are certainly very astonishing indeed ; so
astonishing that I have very great difficulty in
believing them. They shake my faith in the glacial
system altogether ;—or perhaps they ought rather to
shake the faith in Agassiz. They seem to threaten
a *reductio ad absurdum* of the whole theory. If
Brazil was ever covered with glaciers, I can see
no reason why the whole earth should not have been
so. Probably the whole terrestrial globe was once

"one entire and perfect *icicle*." Seriously, — to 1866. answer your questions ;—there is nothing in the least *northern*, nothing that is not characteristically Brazillian, in the flora of the Organ mountains. I did not myself ascend any of the peaks, but Gardner did, and made very rich collections, of which he has given an account in Sir W. Hooker's Journal, and more compendiously in his volume of Travels. The vegetation consists of very curious dwarfish forms of those families and genera which are characteristic of *tropical* America, and especially of Brazil ; together with representatives of some other groups which are widely diffused, but by no means *northern*. So also the vegetation of the table lands has many peculiar forms, but is composed mainly of under-shrubby and herbaceous species, of the same families and genera which in the forests appear as trees and tall climbers.

Certainly, IF Brazil was ever covered with glaciers it seems to me certain that *the whole* of the tropical flora must have come into existence *since*. I also think it clear, on the same IF, that the absence of " glacial action" from Southern Europe must be due to some other cause than climate.

Again, to answer your last question.—Brazil (I speak not merely of the small part which I saw, but of what I have read of, and I have read a good many books of travels in that country), seems to be very deficient in lakes, with the exception of lagoons ("broads" they would be called in Norfolk), on the coast ; of these there are plenty, but they are evidently formed in the same way as the

1866. Norfolk broads, by the natural damming up of the outfall of the abundant waters. Where I travelled, in the higher lands of the interior, the running streams were absolutely innumerable, but scarcely so much as a permanent pond to be seen.

Many thanks to dear Mary for her kind message. With much love to her, believe me ever

Your affectionate friend,

CHARLES J. F. BUNBURY.

I think Joseph Hooker will be as sceptical as myself about the *glaciation* of Brazil.

——— ——

Barton,
February 4th, '66.

My Dear Katharine,

I thank you very much for your kind letter and good wishes on my birthday. Yes, I have attained the venerable age of 57 years, and I feel very thankful, as well I may, for the many blessings which the Almighty has bestowed on me; above all, for the affection of my wife, and of so many excellent and valued friends as I have the happiness to possess. I may well be thankful too for the comparatively good health, and especially for the power of using my eyes in reading as much as I please. But somehow it is my nature, I think, to take more pleasure in looking back than in looking forward: and as I grow older this naturally increases upon me. I look back, certainly, on many faults and follies of my own, but also on a large proportion of peaceful and happy days, and the year (of my life I mean), which has just ended, I look on as a very happy one. I am not at all inclined to

"let the dead past bury its dead";—but much more   1866.
to "trust no future howe'er pleasant."

Have you got *Phascum crispum?* It is not a
common Moss, and I find it here only now and
then at long intervals : but the other day I hit upon
a good patch of it, not far from the spot where I
first found it nearly 40 years ago. I have put up
some of it for you. This mild wet winter is very
favourable for Mosses, and indeed for the vegetable
world generally.

Fanny is well, and we are fagging together at the
classification of our Library Catalogue: the last
step before beginning to write it out (or rather, to
have it written out) in the big book. You would
not easily believe how much labour this classification
requires.

I feel very sorry for Mr. Boxall on account of Mr.
Gibson's death. For Gibson himself, one hardly
ought to say one is sorry (with the exception of his
personal friends) of one who has lived to that
mature age and done his work so well, honoured in
his life and in his death, one can only say that he

> "Home has gone and ta'en his wages."

I have finished Carlyle's "French Revolution,"
and begun the second volume of Stanley's "Jewish
Church."

Pray give my particular love to dear Rosamond,
and thank her very much for me, for her pretty little
present. Much love also to your boys and to Harry.

Believe me ever, your very affectionate Brother,

C. J. F. BUNBURY.

JOURNAL.

1866.   Mr. and Mrs. Abraham arrived.

Tuesday, 6th February.

Walked round the garden and arboretum with Mrs. Abraham. I read some more of Tristram. Sir John Lister Kaye and his daughter arrived, also Mr. and Mrs. Barrington Mills.

Wednesday, 7th, February.

The Miss Waddingtons came to luncheon. Lady Cullum very amusing.

Thursday, 8th February.

Finished reading Tristram's "Land of Israel." It is a good book, but rather diffuse and redundant and heavy.

LETTER.

Barton,
February 8th, '66.

My Dear Lyell,

I have written down on a separate paper (which I enclose) the references to those passages in Tristram's book which have struck me as having more or less of interest with respect to geology. I do not know that any of them are of great importance. Much the most interesting to my

mind is that at p. 575, on the prevalence of fishes 1866.
of *tropical African* types in the Lake of Genesaret,
and the speculation which he founds thereupon,
as to a former physical connexion of the Jordan
valley with the great African system of lakes.
There are some other interesting passages relating
to natural history but not to geology in particular,
such as (at p. 434) the observation that the butterflies in the plain of Genesaret were mostly British
species, "many of which re-appear here after being
supplanted by cognate species in Southern and
Eastern Europe."

Also (what I think would delight Leonard and
Frank), the account of the vultures' nests and the
mode of taking them in the great cliffs near
Hattin, near the lake of Tiberias (p. 446.)

Altogether it is a good book, though it would be
improved, in my opinion, by a good deal of pruning.
According to Sydney Smith's famous classification
of books, it is well-worth borrowing, but perhaps
hardly worth buying.

Much love to Mary.

Believe me ever, your very affectionate Friend,

CHARLES J. F. BUNBURY.

JOURNAL.

Monday, 12th February.

The whole day wasted in a journey to Ipswich to
a meeting on the "Cattle Plague Insurance Act."

LETTER.

My Dearest Cissy,

1866.
    Very many thanks to you for your loving letter and kind wishes on my birthday. I assure you, dear, there is no one whose affection and whose true sympathetic heart I more thoroughly rely than on yours and very few for whose good opinion and good wishes I am more grateful. I have indeed very many causes for thankfulness to the Almighty.

I was very much pleased with your anecdote of Emmy given in your letter to Fanny.

Dear Child, I like that tenderness of conscience, in the unwillingness to accept praise which she thought undeserved.

We got very successfully through our three days of company and our three big dinner parties.

Mrs. Abraham's charming sweetness and Lady Cullum's inexhaustible good nature, good humour and spirits, would make almost any party go off pleasantly.

We had been looking forward to a visit from the Kingsleys this week, but poor dear Mrs. Kingsley is so ill with neuralgia that they are obliged to go to Bournemouth, and we have now no hope of seeing them this season. It is a great disappointment.

Henry will be glad to hear that all the gales of

this winter—the most stormy I ever remember— 1866.
have done no damage of the least consequence in
the arboretum or grounds here, except in one
solitary case. A " Fulham Oak " in the arboretum
was much shattered by the great storm about a
month ago ; but otherwise nothing has suffered, not
even the brittle Catalpa. He will remember that
the Pinus *insignis* here was killed outright by the
Christmas frost in 1860, and I thought we had not
one left ; but the other day to my great surprise I
discovered a fine flourishing young tree of the kind
hidden away in the little family cemetery among the
trees which our Father planted there. I mean to
have it removed as soon as it is safe, to the
arboretum.

I am very glad of William's appointment to the
Board of Military Education, which will bring them
away from Ireland and settle them near London.

The Fenian business is very unpleasant; anything
but a laughing matter in my opinion. In spite of
the many seizures that have been made and the
good success of the trials and even in spite of the
opposition of the priest, the conspirators appear so
numerous and so fanatically resolute, that I am
much afraid there will be an outbreak and much
bloodshed ; helped as the fanatics are by large
supplies of money from the American Irish, and
encouraged by the evident ill-will of the Yankees
towards us.

We shall go up to London in the first days of
March, and we have almost made up our minds to
start for Florence about the middle of the month, to

1866. of plants still living in the Brazilian mountains
see our sisters, hoping to be home again by the end
of May.

Very much love to dear Henry and the children,
and many thanks to him for his kind wishes,

Ever your very affectionate Brother,

CHARLES J. F. BUNBURY.

JOURNAL.

Wednesday, 14th February.

We finished the classification of our library
catalogue.

Thursday, 15th February.

Read the paper by Mr. Lowne, on "The Vegetation of the Western and Southern Shores of the
Dead Sea." Mr. Lowne was one of Mr. Tristram's
fellow travellers, and this paper is especially
interesting in connection with the book I have so
lately read. The results here stated are not quite
in accordance with the conclusions indicated by
Tristram who appears to consider the natural
productions of the lower part of the Jordan valley,
as having a distinctly Indian and tropical character.
According to Mr. Lowne, the affinities of the flora
in question (that of the neighbourhood of the Dead
Sea) are entirely with the floras of Northern Africa,
and especially with the desert floras of Upper
Egypt and Nubia. It will also be found to be
closely related with the flora of Aden in the south
and of the Canary Islands in the west. " I do not
"think" he says "there is anything to warrant the

"ordinary belief that the flora of the south of the   1866.
"Ghor," is an Indian type."

Kingsley arrived. I had a pleasant little walk
with him.

Mr. Abraham dined with us:—very pleasant
conversation with him and Kingsley.

A beautiful day. Strolled with Kingsley. A
very pleasant evening of conversation with Kingsley.

Kingsley went away early. Visit from Arthur
Hervey. Read in the *Times* the debate on Ireland.

I wrote by return of post a long and rather
impetuous letter to Lyell, expressing my scepticism
on Agassiz's "Researches on Glacial Phenomena."
This day, he has sent me two letters to read from
Joseph Hooker and Charles Darwin, to which he
had communicated both Agassiz's remarks and my
letter. Hooker is nearly as incredulous as myself.
Darwin believes in Agassiz's observations, and
speculates on the possibility of glaciers having
formerly extended from the Andes to the Brazilian
mountains, when these latter were much higher
than now, and directly connected with the Andes,
in fact a branch of them (at least so I understand
him). He also thinks that there are certain genera

which indicate former connection with colder regions. This I was at first inclined to deny altogether, the rather as those genera which Darwin mentioned as instances certainly did not support his doctrine, being widely diffused genera of wide climatal range, but I afterwards recollected some instances of groups of closely allied species (groups characteristically South American) which are common to the Andes and the Brazillian mountains, and may with some plausibility be conjectured to have spread from the former to the latter. Such are Gaultheria Gaylussacio, Fuchsias more particularly Drimys. I accordingly write to Lyell to this effect.

Wednesday 21st February.

A very cold day. Attended Gaol committee at Bury. Met in the town — Colonel Anstruther, Barnardistons and Rowley. We dined at Lady Cullum's, and met Sir John Walsham and Sir James Simpson.

LETTER.

Barton,
February 20th, 1866.

My Dear Lyell,

Very many thanks for sending me Hooker's and Darwin's letters, which I have read with great interest. I agree in almost everything that Hooker says, as far as I can make him out, but his letter is *very* hard to read. I differ from Darwin as to the

plants which he quotes, as instances of the occur- 1866.
rence of temperate forms on the Organ mountains ;
he seems to consider as a "temperate" genus every
genus which is found *at all* in temperate climates,
and here I think him mistaken. I think I mentioned in my former letter, that, besides the strictly
tropical forms on those peaks, there are species of
genera which are very widely spread, and not
specially either tropical or the reverse. Such a
genus is Hypericum, one of those which Darwin
enumerates ; it is found in almost all parts of the
world, *except* very cold countries. Clematis (which
he does not mention) is another instance of the
same kind Drosera and Habernaria (as Hooker
points out) have certainly their maximum within the
tropics. If there are Vacciniums on the Organ
mountains, they are of the sub-genus (Gaylussacia
of Humboldt), which belongs specially to South
America, and of which there is a species even on
the coast of Brazil, in the island of St. Catherine.

If the Brazilian mountains were once a branch of
the Andes (which I infer is Darwin's notion) I
should have expected a greater number of the
peculiar characteristic* forms of the Upper Andes
to be found on the mountains of Minus, &c., such as
those "Rhododendrons of the Andes" (Befarius),
of which Humboldt talks so much. There are
some such : Gaultherias, Gaylussacias, Escallonias,
&c., but not so many as one would expect. The
strongest case, perhaps, in favour of Darwin's view

---

* Charles Bunbury is not quite certain whether to call them characteristic or
peculiar forms. I think the former.—F. J. B.

K

1866. and against mine, is the genus Drimys (the Winter
Bark). Whether the American forms of Drimys be
all varieties of one species, or a group of closely
allied species, they certainly afford a most striking
instance of a group of very near relations ranging
along the Andes, from Cape Horn all through South
America into Mexico, and re-appearing conspicuously on the table land of Brazil. I do not
know whether they are found anywhere *between*
Minas and the Andes. It is certainly quite allowable for Darwin to say, that they must have
migrated to the Brazilian uplands when these were
more closely connected with the Andes than they
now are. Fuchsia comes nearly into the same
category with Drimys, except that there is a greater
variety of forms, and some of them more decidedly
distinct. I doubt whether either Fuchsia or Drimys
is found very high up on the Andes.

I acknowledge that, in my former letter, I did not
sufficiently consider the possibility of the Organ
mountains and those of Minas having been formerly
much higher than now, and of their upper regions
having been *"glaciated"* while in that position.
But after all, as Hooker says, the information in
Madame Agassiz's letter is almost too vague to
afford any safe ground for fighting upon. I think
the meaning *must* be, that the "glacial" marks were
observed *down to* (not *up* to) 3000 feet. This is a
most material point.

I do not agree with Darwin, that the nature of the
vegetation of New Zealand, gives us reason to
believe that tropical families of plants could bear a

cold climate. However luxuriant the vegetation of 1866.
New Zealand, it does not, I think, include any
really tropical types. I am not so sure, however,
about Chiloe and Valdivia.

I am very glad to hear that Darwin's health is
better.

Believe me ever your very affectionate Friend,

CHARLES J. F. BUNBURY.

You may perhaps perceive that a certain degree
of change has come over me while I have been
writing this letter. I feel that I was perhaps too
absolute in my first incredulity as to the possibility
of glaciers on the Brazilian mountains : though I
still think it quite incredible *if they had only their
present elevation* ; and I have recollected (as I
noticed above) some instances of Brazilian plants
which might with some plausibility be supposed to
have migrated from the Andes.

JOURNAL.

Saturday, 24th February.

Kingsley arrived about mid-day. I went with
him after luncheon to Ickworth, and saw the Arthur
Herveys. Lady Cullum dined with us. Kingsley
very agreeable.

Finished lecture 30 of Stanley's "Jewish Church."
The history of Elijah is magnificently told in this
chapter ; the narrative of his contest with the
prophets of Baal on Carmel, and the reflections on

1866. his loneliness in the desert of Horeb, are wonderfully noble and impressive.

Sunday 25th February.

We went to morning Church and received the
Sacrament.   I went to afternoon Church with
Kingsley who preached.   A very agreeable evening
with him.   He is now I am happy to find in much
better health than when he was with us in
September.   I wish I could remember more of his
conversation.   He is more and more an admirer of
Darwin's theory of variation and natural selection;
thinks it becomes more and more evident how much
more "living" and "fruitful" this doctrine is than
any previous one; how many more phenomena it
explains and how much more fruitful it is in
interesting results.

He thinks Alban Butler's "Lives of the Saints" a
worthless book—almost dishonest.   The old legends
are so trimmed and diluted and tamed down to suit
the taste of good society, and of the times in which
he wrote, that everything characteristic is cast
out from them, and they retain nothing of the old
spirit, and give no idea of what was really believed
in "the ages of faith."

Kingsley told me in a most delightful way a
Cornish legend about two saints of that country who
were also giants—St. Kevern and St. Just.   It was
as good as one of Crofton Croker's Irish legends,
but all depended on the language and manner of
telling it; so I cannot repeat it.

He (Kingsley) has a strong antipathy to monas-

ticism.   He admits indeed that it was better than
the uncontrolled reign of physical force, but he
contends that the false views of human duty and
human nature which were continually taught by the
monks, retarded by centuries the progress of
mankind. Their fatal errors were the depreciation of
marriage and family life, and their degrading
estimate of woman.   Instead he says of trying to
make use of the elements of law and order which
did exist around them, instead of trying to induce
men to live in society with some kind of decency
and justice, they laboured to set up an entirely
strange and unnatural order of society founded on
the reverence for celibacy and asceticism.   I cannot
do any justice to his force of expression.

He thinks that this belief in the surpassing
excellence of celibacy, and the ascetic doctrine
generally as well as many others of the characteristic
practices and doctrines of mediæval Christianity
(beginning even from the third century) were
derived originally from the Buddhists of Eastern
Asia.   Probably they came through the medium of
the Gnostic sects; but he says he knew really
nothing about the Gnostics.   He has read all that
is now to be found concerning the Gnostics, and his
conclusion is that all the information we possess
about their doctrines comes from their enemies, and
is not to be trusted; even St. Augustine, one of the
greatest geniuses that ever lived, was quite capable
he thinks, of misrepresenting his opponents.

Kingsley says of Carlyle's " French Revolution,"
that he learned more about the real nature and the

1866. real meaning of the French Revolution from that book, than from the whole of what he has read besides about it. Carlyle alone made him understand the Revolution ; all the other works have only filled in the outlines of the grand sketch.

Kingsley said that Barbadoes is the most flourishing of our West Indian Colonies, because it contained the *smallest* amount of unoccupied land on which the Negroes could squat, and Jamaica the least prosperous because it contained the greatest quantity of such land.

We talked of "Spiritualism" and agreed that those who say that such exhibitions confirm their faith in a future state, only acknowledge thereby the exceeding weakness of their faith.

Monday, 26th February.

Kingsley went away early.

My Barton audit—very satisfactory and comfortable.

Tuesday, 27th February.

Read Stanley's 32nd lecture—"The History of Jehu." I think Stanley's judgement of this tyrant fully indulgent enough, to say no more. Jehu appears to me to have been a thorough Oriental usurper, just such as one reads of continually in the histories of Persia and the Mogul Empire ; a man without faith, without scruples and without mercy ; whose religion showed itself only in savage bigotry.

An allusion in Stanley led me to read again that magnificent paper on "Paradise Lost," in which Milton describes as devils, the various false gods of the nations connected with Palestine.

We left our dear home and came up to London. Found everything in proper order at 48, Eaton Place.

I went to the Kensington Museum and spent an hour there; saw many interesting things. Clement came to luncheon.

Visited Lady Grey—pleasant. Saw Sir Edward Head, Gurney Hoare and Mr. Boxall at the Athenæum.

A terribly cold day.

We visited Norah Bruce, saw her and Catty. Dined with the Charles Lyells.

Yesterday Sir Frederick Grey told me that he had just received a very interesting account of the eruption of Santorin from a Captain of a ship, and had sent it to Charles Lyell. He said it had been known for some time in the navy—that if a ship lay at anchor within the ancient crater of Santorin, her copper would soon be effectually cleaned by the peculiar quality of the sea water; doubtless owing to the emission of some acid gas from the bottom.

1866.   Sunday, 11th March.

We went to morning Church (St Michael's Church, Chester Square) and heard a good sermon from Mr. Hamilton.   Sir William Gibson Craig called.   We visited Katharine and Mary.   I looked over minerals with Leonard.

———

Monday, 12th March.

Edward dined with us, and brought us the latest information as to the Gladstone Reform Bill.

———

Tuesday, 13th March.

Read the debate on the Reform Bill.   Sent some money to Mr. West* for the sufferers by the cattle plague in Cheshire.   The MacMurdos dined with us—very pleasant.

———

Wednesday, 14th March.

Wrote to Scott.

We dined with the Bruces.   *Almost*, but not quite a family party—afterwards went to an evening party at Miss Richardson's.

———

Thursday, 15th March.

Went to two picture-cleaners.   Fanny's afternoon party—very pleasant.

———

Friday, 16th March.

We dined with the Charles Baring Youngs.

———

Saturday, 17th March.

Read part of Bolingbroke's letter to Sir William Wyndham.

* Incumbent of Stanney.

Sarah Craig came to luncheon with us,—Katha- 1866.
rine and Harry Lyell and their three boys dined
with us.

————

Monday, 19th March.

Visited Sir John Boileau and had a long and
pleasant talk with him—also the Benthams—also
the Gibson Craigs.

Finished reading Bolingbroke's letter to Sir
William Wyndham ; certainly a very fine specimen
of political writing. The apparent object of it is
to vindicate his own political conduct : first, during
the last four years of Anne's reign, while he was
acting (as he represents the case) as a true constitutional Tory, and not a Jacobite ;—secondly,
during the early part of his exile in France, when
he was Secretary of State to the Pretender ;—
thirdly, after he had been dismissed from that post,
and had again separated himself from the Jacobites.
In the first of these periods, he is extremely bitter
against Harley, to whom mainly he attributes the
downfall of the Tories. His positive assertion, that
neither the party in general, nor himself in particular
had any correspondence at that time with the
Pretender, nor any intention of bringing him in, has,
I fancy, been contradicted by documents since
brought to light. In the second part he does not
seem to explain very satisfactorily why he entered
so suddenly into the Pretender's service. But his
picture of the miserable intrigues and pettinesses
and follies of that exiled Court, is very vigourously
drawn, and no doubt very true.

The most powerful and striking part of the whole work is the last,—in which he displays the wretched bigotry of the Pretender; and the dangers to which the British Constitution would have been exposed in case of his restoration. Whether in this he may not justly lay himself open to a charge of treachery may be a question. Altogether, I think Bolingbroke throughout this essay much more successful in damaging others than in vindicating himself. His remarks on the intolerance of Christians towards one another are very severe and lamentably true.

Wednesday, March 21st.

Began to read Bolingbroke's "Dissertations upon Parties." The dedication to Sir Robert Walpole is keenly sarcastic.

Thursday, 22nd March.

Scott arrived from Barton and we had a talk with him.

Friday, 23rd March.

Visit from Susan MacMurdo. Charles and Mary dined with us—very pleasant.

Monday, 26th March,

Visit from Sir Edward Head. We dined with the Boileaus—very pleasant.

Tuesday, 27th March.

Charles and Mary called to take leave of us, as they were going to Weymouth.

Went to the Zoological Gardens. We had a dinner party, but Fanny was ill and not able to appear. We had the MacMurdos, Matthew Arnolds and Louis Mallets.

Saturday, 31st March.

Called on Sir John Bell. Also on the Adairs.

April 3rd.

William Napier dined with us ; very pleasant.

April 4th, Wednesday.

Went to Fulham and saw the MacMurdos in their new house. They were very pleasant. Then went to the South Kensington Museum.

April 5th.

Fanny able to go out for the first time since March 26th. We drove to Putney and spent the afternoon with Emily and her children.

10th April, Tuesday.

From the 23rd March to the 9th April when we left London, next to nothing was done by me in study ; a very idle, useless and unsatisfactory time, partly owing to Fanny's illness and the uncertainty of our departure.

I saw a few very pleasant people — the Evans Lombes, who dined with me on the 28th ; the Boileaus, Bowyers, Arnolds, MacMurdos and William

1866. Napiers. I visited the Horticultural Gardens, the
Zoological Gardens (where I saw the curious new
seal or sea bear) and the South Kensington
Museum.

We left London on Monday, the 9th April and
went down, we two and Kate MacMurdo, the two
maids (Bessie and Julia), and my valet (Edgar) to
Folkestone, where we slept. We had a pleasant
visit from the Miss Richardsons in the morning, and
crossed to Boulogne.

Wednesday, 11th April.

We left Boulogne and arrived at Paris a little
before six, at the Hôtel Brighton, Rue Rivoli,
looking over the Tuileries Gardens.

Thursday, 12th April.

Visited Fanny's Aunt, Madame Byrne and Mrs.
Power, and strolled about Paris all day, amusing
myself with the gay shops and street sights.

Friday, 13th April.

We called on Madame Tourgueneff — very
pleasant. We dined with Mrs. Power and Mrs.
Byrne.

Saturday, 14th April.

We spent much of the day in the Louvre. The
galleries very magnificent, and the pictures shown
to the best advantage. We went to an evening
party at Madame Tourgueneff's.

A beautiful summer day. Very pleasant drive to the beautiful Bois de Boulogne.

A small evening party at Fanny's Aunt's. Talk with Mr. Tourgueneff and Colonel Danner. Mr. Tourgueneff's opinion of the King of Prussia is that he does not anticipate war, but follows blindly the counsels of Bismarck, without at all forseeing the consequences of the step he is about to take.

———

From Paris to Dijon (6½ hours) by express train. As we approached Dijon there is a long succession of tunnels through the limestone rock; the first of them of very great length. In this high country we passed Montbard, where Buffon lived.

———

We walked about Dijon in the morning; a fine and curious old town, retaining much of the old French character. The Museum in the Hôtel de Ville contains a rich collection of art and antiquities very remarkable for a provincial town. The most striking are the huge, gorgeous and most elaborately ornamented tombs of two Dukes of Burgundy, Philippe le Hardi and Jean Sans Peur.

We then visited the old Church of Nôtre Dame; a most remarkable front, rising (as in some of the old Churches at Lucca, I think) to an extravagant height above the body of the Church, and adorned with a double range of arcades on tall slender pillars.

1866. We saw also the Cathedral, more curious than beautiful.

We started from Lyons by the 12.35 train, but Fanny being taken ill, we were obliged to stop at Mâcon.

LETTER.

Mâcon (sur Saône) not Mâcon U. S.
April 19th, 66.

My Dear Mary,

I shall probably not send off this letter for some time, but I have a mind to surprise you by its date, for probably neither you nor any of our friends will have guessed that we spent this last night at Mâcon.

We have not met with any accident, but poor Fanny was attacked with such a dreadful headache in yesterday's journey, that she was unable to go on, and we had to halt here. We have been very lucky in finding a good inn and very spacious, airy, handsome rooms, with an extremely pleasant look-out over the quays and the fine broad stream of the Saône. And here we have determined to remain to-day, that Fanny may have time to rally completely. If it had not been for this unfortunate headache, we should have had a very pleasant journey from Paris; the weather has been and is glorious,—to-day is like a June day, and the country is in its very best looks, with the tender varied colouring of the young leaves and grass and the profuse blossoms of the fruit trees. We hope

to reach Lyons tomorrow, and Florence, *quand il* 1866.
plaira à dieu. Travelling has done me a world of
good : I feel wonderfully better than when I was in
London. Kate MacMurdo is a charming companion.

(Nice, April 26th.) We have arrived safe and
sound so far on our way and Fanny happily has had
no return of headache. At Lyons I had a glimpse
of the Museum of Natural History, and saw by far
the most magnificent suite of specimens of the blue
and green carbonates of copper from the mines of
Chessey that I ever set eyes on. Railway travelling
is not with me a favourite mode, and the arrangement of the French railways (at the stations, I
mean) I think detestable : but I must acknowledge
that I have not lost so much of the beauty of the
country as I expected. In particular, the valley of
the Rhone appeared much more beautiful than I
had ever thought it before, from the variety of
bright delicate tints of the young crops and trees.
I have always thought it very interesting to observe
the successive appearance of southern forms of
vegetation as we travel from Paris to the Mediterranean ; and in this tour we have had the additional
pleasure of observing the different stages of progress
of the same kinds of plants in different latitudes.
For instance, between Paris and Dijon the Lombardy Poplars were still leafless ; about Mâcon
they were clothed with their pale yellowish young
leaves, and in the valley of the Rhone they were in
full-green foliage. In that same valley we first saw
Fig and Mulberry trees in leaf, and the vines

1866. beginning to put out their leaves. From Toulon
hither was a most beautiful journey, especially
beween Frejus and Cannes, where the railway skirts
the coast, which is like a piece of the Genoese
Riviera. But it is very tantalizing to a botanist to
be whisked at such a rate past such a profusion of
beautiful flowers,—such tufts of Cistuses, both rose
coloured and white, and such masses of the rich
purple crested Lavender (Lavandula Stœchas).
I hope, however, to be able to gather some plants
in the vetturino journey from hence to Genoa.
Certainly botanists were never intended to travel by
railway. Here we seem to be in the middle of
summer, or almost within the tropics. The gardens
full of roses and geraniums in full blossom, and
bouquets of orange flowers sold about the streets.

(April 28th.) The weather is most lovely, the
situation of this hotel excellent, and the view of the
bay delightful. But we have not on the whole
spent our time at Nice delightfully. I leave Fanny
to give you the account of our adventures here.
We shall not be able to get away before Monday
the 30th. But we are fortunate to be in such good
quarters.

After I wrote the last sentence we had a very
pleasant drive, first to Villefranche and then up the
charming valley of St. André. We were often
reminded of Madeira and Teneriffe, especially as we
came back, when the evening air was loaded with
the scent of the orange blossoms.

Ever your affectionate Brother,

CHARLES J. F. BUNBURY.

JOURNAL.

Nice,
Thursday and Friday, April 26th-27th.

1866.

In the Public Gardens near the sea, besides a great number of Date Palms, I see some large trees of the Ombü of Buenos Ayres, the Phytolacca dioica, also Schinus Molle, forming small trees of very graceful character ; Pittosporum tobira loaded with blossom, Osteospermum mæniliferum from the Cape, and some Australian Acacias and Eucalyptuses in flourishing condition.

Saturday, 28th April.

We were detained in Nice by the illness of Fanny's young maid Julia.

Sunday, 29th April.

We visited Dr. Pantaleono. He made arrangements for us with the vetturino.

Monday, 30th April.

We saw the telegram with the account of the decision on the Reform Bill. We started from Nice about 10.45 a.m., in a very comfortable vetturino carriage with four horses, and arrived at Mentone about 2.20. A most beautiful journey, but the day, unfortunately, was not favourable. In the earlier part of it indeed, in the very long ascent from Nice, we enjoyed most beautiful views of the valley of the

L

1866. Paglione and those rivers that run into it, the bay of Nice, the town, the olive-clad hills with their villas and convents.

On the ascent from the valley of Nice, a great many beautiful wild flowers on the banks and rocks by the sides of the road ; noticed, in particular, the lovely Convolvulus althæoides and a fine Cytisus (or Genista) very prickly, but covered with bright yellow blossom.   Olive and Fig trees seen even above Turbia, certainly near 2000 feet of elevation, but sparingly, and the Fig only now coming into leaf.   Near the same elevation I saw a deciduous Oak (probably Quercus pubescens) covered with young leaves.   Euphorbia dendroides in great profusion all over the high arid limestone country about Turbia, forming dense, hemispherical bushes covered with yellow flowers.

————

Tuesday, 1st May.

Pont St. Louis, exactly at the present French and Italian frontier, an old bridge over a torrent, in a very fine deep rocky ravine.

The coast between Mentone and Ventimiglia very fine.

Bordighera : the multitude of Date Palms cultivated here and along the coast on to San Remo, very remarkable ; and here and there whole groves of them of various ages intermixed, have a very striking and exotic effect — quite an African character.   But the beauty of most of the individual trees is much marred by the way in which they are treated—the leaves being either all tied up close, or the outer ones cut off, so as to mutilate the crown and deprive it of its full expanse. This is to furnish "Palm branches" for Rome.

Convolvulus althæoides forms lovely wreaths of flowers on the rocky banks and the stone walls of the vineyards. Gladiolus communis and the Borage very abundant everywhere among the young corn and in the Olive grounds, and very beautiful. A very pretty little Silene with small glaucous leaves and delicate blush-coloured flowers, growing plentifully on the sands of the sea-shore, near Bordighera. Great abundance and variety of Euphorbias on the rocky and sands shores.

San Remo, a beautifully picturesque town. A very nice new hotel outside the town, at the foot of the hill. The beautiful orchards of Lemon and Orange trees, especially the former, continue very conspicuous objects along this part of the coast.

From San Remo to Oneglia, the coast less interesting, except the quaint and picturesque little old towns, through some of which we passed, through streets so wonderfully narrow, that the carriage seemed in constant danger of collision with the walls of the houses.

Porto Maurizio a larger and more important place, exceedingly picturesque, and a fine town with well flagged streets.

———

Wednesday, 2nd May.

Cape delle Mele — formidable precipices — the road would be improved by a good parapet. Rocks

1866. of black shaly limestone curiously variegated with a network of white veins.

A long succession of bright, white towns along the coast—marvellously picturesque; some on the shore, some at various heights on the olive-covered hills.

The cultivation of this coast shows marvellous industry; the hillsides where they appear almost precipitous, everywhere terraced and planted with Olives which grow magnificently. The huge, fantastic, knotty and twisted trunks of the old Olive trees, and their light, ashy-grey, misty looking foliage are very interesting objects.

The vegetation in general of the same character as the two last days, and we did not add many specimens to our botanical collections: only Polygala major, with fine deep rose-coloured flowers; a beautiful Cistus (? incanus) in one place; and on the marble rocks of the Cape of Noli, a curious little plant (in seed) which I did not recognize* I forget whether it was on the rocks of Cape delle Mele, or of the next headland, that I saw the Oleander growing wild, but out of flower. The leaves of Pancratium maritimum (I suppose) are seen in plenty on the sea-side sands in more than one place, particularly between Albenga and Finale.

Cape di Noli; grand precipices of very hard and compact limestone, real marble, in parts black, in others reddish. A heavy surf breaking finely on rocks here, and on the other projections of the coast.

* It was Coris Monspeliensis.

3rd May, 1866.

We remained this day, quiet, at Savona.   In the afternoon we enjoyed a pleasant drive up a very pretty valley, or rather glen, our road being parallel to the course of a new line of railway to Acqui and other places in Piedmont.   This glen repeatedly reminded me of some in Wales, though the different character of the vegetation, and the terraced sides of the hills made great differences.

Found a lovely little Saxifrage in flower, growing in plenty on wet dripping rocks ; very like Saxifraga *cuneifolia*,* but much smaller and more delicate.

LETTER.

Savona,
May, 3rd, 1866.

My Dear Henry,

I wrote to Cissy from Avignon, where we halted the Sunday before last ; and I believe Fanny has written to her since.   You will perceive by the date of this letter that we have not made our journey so expeditiously as we had calculated upon ; but we have gone I think, as fast as is pleasant and as fast as suits either Fanny's health or mine.   At Nice we had some botheration about a vetturino, and were moreover delayed a day by sudden illness of Julia. But on Monday the 30th we set off on our vetturino journey along the Riviera to Genoa, and we should have had a very pleasant journey, if the weather had not (until to day) been pertinaciously wet, lowering and cold ; a very unpleasant surprise at this season, and in this country.

* Certainly a state of cuneifolia.

1866.   We are now halting here for a day, as Fanny
is tired, but to-morrow we hope to get on to
Genoa.

This is now the third time I have travelled along
the Riviera, and it appears to me as beautiful and
interesting as ever.   Though the bad weather we
have met with has been unfavourable to the scenery,
I have enjoyed the charming vegetation perhaps
even more than before, as so many beautiful things
are in blossom; nor do I think I was ever before so
much struck with the rich effect of the Orange and
Lemon orchards, which are perfectly loaded both
with fruit and flowers.

The road is certainly immensely improved since
you and I first travelled it in 1827, and I think it is
improved even since 1848; in particular almost all the
torrents are now bridged, which is a great gain.
But a more efficient parapet wall in several places
would still be a great improvement to the road;
there are parts of it, round the faces of the head-
lands, which are still rather alarming.

Those who shall travel by the railway, when it is
finished, will travel mostly in the dark; it is a string
of tunnels, of which we could see the mouths as we
passed along.   The inns as well as the road are
improved since my first acquaintance with the
Riviera.   But the look and dress of the people,
their houses, their modes of cultivation, the look of
their quaint picturesque towns, and the smell of the
streets, as well as of course, the aspects of nature,
are unchanged since we saw this country in 1827;
and long may they continue.

(Genoa, May 6th).   We arrived here the day 1866.
before yesterday, and unexpectedly find ourselves in
the midst—not indeed of actual war—but something
more than rumours of war.   The excitement is
immense, preparations for war most active ; the
enrolling and mustering and marching of troops
going on incessantly ; the ordinary passenger traffic
on the railway northward from this has been
stopped, except two trains by night, in order that
the railways may be entirely devoted to the con-
veyance of troops.

It is a strange fatality that Fanny and I who
are ourselves neither warlike nor revolutionary,
should fall in with war and revolutions every time
we come to Italy.   It will be very unlucky if we are
a second time prevented by war from seeing Venice
and Lombardy, but there is every appearance of it.

Still I am not entirely without hopes that the
cloud may blow over ; I cannot yet bring myself to
believe that Prussia and Austria will be so madly
wicked as to go to war ; and I hope Italy will not be
so insane as to go to war single handed.

We yesterday visited our mother's grave, and we
have made some arrangements for making it a little
more cheerful (it is gloomily overgrown with trees
and shrubs), and for having it tended.

We hope to start to-morrow, by vetturino, for
La Spezia, from whence there is now a railway to
Florence, so that we may expect (D.V.) to arrive at
our journey's end on Wednesday.   I am in much
better health than when in London, and Fanny is I
think, well on the whole.

1866.   Much love to dear Cissy and all the darling
children.

Ever your very affectionate Brother.

CHARLES J. F. BUNBURY.

JOURNAL.

Genoa.
Saturday, 5th May.

Great excitement. Active preparations for war :
the city full of troops—incessant arrivals, mustering
and marching of soldiers. The railway to Alessandria
re-taken for mustering of troops, and ordinary
passengers' traffic stopped except by night. A
recent decree of the Government also gives com-
pulsory currency to the paper money.

Beautiful day. We visited three palaces. Then
we went to the English burial ground at San
Benigno, and saw my mother's tomb. On descending from thence we had a glorious view of the whole
of Genoa and the mountains to the East, ending in
the fine headland of Porto Fino.

Monday, 7th May.

We travelled from Genoa to Sestri di Levante,
and on the way we met the Wedgwoods.*

The day very sultry but favourable to the enjoyment of the scenery. It is a most beautiful journey
of which I had a vivid impression from our
experience in 1848 : and it strikes me now, just as it
did then, as being at least equal, if not superior in

* Mrs. Wedgwood the daughter of Sir James Mackintosh.

beauty, to any part of the western Riviera. At 1866. Sestri, behind our hotel, an avenue of Orange trees, many in full blossom, perfuming the air deliciously.

The hotel at Sestri, the same in which we spent such a delightful day in the summer of 1848, with the lovely little wooded peninsula full in view, and the grand panorama of mountains behind.

————

Sistri di Levante to La Spezia.

The most conspicuous wild plants are two or more species of Broom, brilliant with a profusion of golden yellow blossom, one of them the very thorny bush cytisus spinosus (?) which we have seen in abundance along both the Riviera, all the way from Nice, —another, a Genista (? Genista pilosa), of lower growth and without thorns ; a heath of which the blossoms are just past (Erica arborea, I think); and the beautiful white-flowered Cistus (salvifolius, probably). I saw also a little of Lavandula stoechas, but no rose-coloured Cistus. The Pines which clothe some part of these mountains, but not the highest parts, are all (as far as I saw) pinaster, not Halepensis. Our common Furze, Ulex, in abundance on the descent of the mountains towards Borghetto, and thence eastward. It is not a common plant in the country we have travelled through.

On one of the last hills we crossed, we find Serapias lingua, and *Ophrys* aranifera, growing amidst the grass.

1866.   We spent the evening with the Somervilles, who
I had not seen for 23 years.

We remained quiet at La Spezia ; we dined at
the Somervilles, and spent a very pleasant evening
with them. Mrs. Somerville, notwithstanding her
great age, is wonderfully active-minded, full of
interest and information about the newest scientific
researches, and her intellect as clear and lively
as ever.

The day splendid and very hot, but with a fine
sea breeze. We went out on the beautiful gulf for
an hour and enjoyed the lovely scenery.

The Carrara mountains looked most beautiful
towards evening, with the clouds resting on their
peaks.

La Spezia itself, quite altered since I was last
there, and quite spoilt for visitors and travellers, by
the extensive constructions connected with the great
naval establishments of the government, and by the
works for the railways.

Miss Somerville gave me a specimen of the Pteris
Cretica, which she had gathered near La Spezia.

Another beautiful and very hot day.

In the afternoon we took an omnibus from the
hotel, and in company with the two Miss Somervilles went to Porto Venere. The road dreadfully
rough, owing in great measure to the government

works; we were worse jolted than I remember to 1866. have been since the old Cape days, but the views beautiful. At Porto Venere we left the carriage, and entering by the old gateway through the old walls, walked through this singularly quaint and curious little old town. It is marvellously picturesque, with its old houses piled steeply up in almost a pyramidal form to the top of a bold hill, and its grey battlemented walls running up the almost precipitous ascent. The street very narrow, impassable for carriages, well paved, but as dirty as if it had never been cleaned since the middle ages. The inhabitants appear a fine set of people. We walked to the ruined fort on the extremity of the headland beyond the town, and looked on the cliffs to the westward and on those of the island of Palmaria opposite. Very fine bold cliffs of very hard limestone, going sheer down into beautifully clear blue water, and hollowed at their bases into caverns by the waves. Miss Somerville told me that in clear weather she has seen from this headland the shipping, both at Genoa and Livorno.—

On the island of Palmaria there is a large convict prison. A very large fort on the top of the island, and other fortifications on all the prominent points on this side of the bay.

In the rocks of the headland just outside the town, we observed good examples of the fine Porto Venere marble. The principal quarries of it are on the island.

I had another talk with Mrs. Somerville in the evening.

                          May 11th. Friday.

We left Spezia very early by railway, and arrived at Florence before noon. The day splendid and very hot.

The first part of the journey from La Spezia to Pisa was very pleasant; we were alone in the carriage, the speed was moderate and the stoppages so frequent and in such convenient spots, that we saw the country very well. Glorious views of the Carrara mountains; they are certainly one of the finest mountain groups I have ever seen; and their crags and furrowed precipices showed with the finest effect in the morning light

Between Sarzana and Pietra Santa, the railway runs in general nearly parallel to the old road, nearly skirting the bases of the hills; and I was struck as formerly by the singular abruptness with which the rocky limestone hills rise from the plain, almost in cliffs, without any intermediate gradations —the plain (alluvial, I suppose), stretching away from their base to the sea, smooth and unbroken.

——————

Monday, May 14th.

We went to the Senate with an order from the Marchese Cappione—heard nothing distinctly.

We removed from the Hotel d'Italie (which is very good) into furnished apartments. Via di Seraigli.

——————

Tuesday, 15th May.

We went to the Palazzo Pitti. Spent most of the morning in the enjoyment of that exquisite collection of pictures.

Spent two hours delightfully at the Uffizii gallery. The Medici Venus.

Edward arrived.   Began to read Trollope's "History of Florence."

We walked home through the Boboli gardens. We called on Professor Parlatore, with whom I had some botanical talk.

Walked in the Boboli Gardens with Kate. We went with Susan and Joanna to Fiesole—to the Church on the top of the hill; a very agreeable drive, the day being fine after several bad ones we enjoyed it the more.

The view of Florence and the valley and the surrounding hills spangled with white houses, from Fiesole most beautiful.

We passed the Villa Mozzi, formerly belonging to Lorenzo di Medici, the view from which is so admirably described by Hallam.

It belonged at one time to Lady Orford, the daughter-in-law of Sir Robert Walpole. It now belongs to Mr. Spence.

This being a festa, the Piazza and Cathedral of Fiesole were crowded with people.

In the evening we saw in the garden of our villino, the first fire-flies.

Wednesday, May 23rd.

A windy and bitterly cold day. We went to the Accademia delle Belle Arti; an interesting and valuable collection of pictures, by the early masters, some of them admirable in themselves, and all instructive and valuable for the history of art; with this object they are very well arranged chronologically, so as to show the progress from the rude Byzantine type up to Pietrie Perugino and del Sarto. Of Andrea, some especially fine works here.

The day so windy and so bitterly cold, that we went nowhere else. Susan and Joanna's second evening party: many pleasant people. I had much conversation with Baron Carlo Poerio (the famous Neapolitan patriot, a very mild, pleasing man), M. Ernest de Bunsen and Marchese Franzoni. Baron Poerio much interested in English politics—has known most of the eminent English politicians of our time—has a prodigious admiration for Mr. Gladstone—rather despondent about Italian politics, especially finance—Italian representatives running too much into radicalism. The present Chamber of Deputies containing too large a proportion of new and inexperienced men, and therefore apt to be rash.

Marchese Franzoni seems a very learned man, much devoted to classical studies.

————

Thursday, May 24th.

Confined to the house by a cold. Read Trollope's "History of Florence."

Friday, May 25th.   1866.

A change of weather. Rain and a soft damp
relaxing scirocco. Confined to the house all day.
Susan and Joanna supped with us. Met Madame
Ernest de Bunsen, and M. and Madame Charles de
Bunsen came in the evening. We had music.

LETTER.

Villino delle Torce,
Florence, May 25th, 66,

My Dear Katharine,

The weather has been very much against
us ; the first two or three days after our arrival were
delightful, but since then it has been most persever-
ingly bad—sometimes wet, sometimes a cutting,
cold wind like March, altogether as unlike as possible
to what I had believed an Italian May to be.

We had a delightful journey along the Riviera,
especially the two days from Genoa to La Spezia
(for in the early part we met with much bad weather)
and spent two days very pleasantly at La Spezia,
seeing much of the Somervilles.

Mrs. Somerville is really wonderful, her mind as
clear and active as possible, as much alive as ever
to all that is going on in the scientific world,
and her conversation charming.

I had seen the Genoese Riviera twice before,
but never in spring, and I thought it more beautiful
than ever. The varied greens of the young foliage
of the Mulberries, Figs and Vines, and of the grass
and young corn, combined with the grey of the

1866. Olives, heightened the beauty of the scenery and the profusion of lovely wild flowers made it like a continued garden.

I do not think that we met with anything rare but we collected and dried a good many plants in that part of our journey. The vineyards and Olive-grounds were full of beautiful "weeds," in particular Gladiolus communis, Muscari comosum, Allium roseum and Allium triquetrum; and on the banks and walls bounding them a lovely Convolvulus (althæoides) with large rose-coloured blossoms. The wild, rough parts of the hills under the pines were gay with various kinds of Genista and Cytisus, loaded with blossoms, a rose-coloured Cistus and a white one (albidus and salvifolius), and a very fine Lavender, Lavandula Stœchas.

Of Ferns we saw hardly any except the Adiantum Cap. Ven. and Asplenium ceterach, and near La Spezia, Selaginella denticulata. Martha Somerville gave me a specimen of Pteris Cretica which she had gathered in that neighbourhood; Mr. Ball, whom we have met here, says it is plentiful in the glens of the Carrara mountains. Prof. Parlatore, whom I have seen once, inquired after you; I have no doubt he would have shewn me every attention and I had hoped to see the Botanic Garden and the Herbarium under his auspices; but he has been ill from the same cause which has confined me also to the house. Susan and Joanna are as kind and pleasant as possible, and I have met at their house several agreeable and interesting people, snch as Baron Poerio, Marchese Franzoni, M. and Mme.

Ernest de Bunsen, Mr. and Mrs. Ball, M. Pulsky, 1866. Mr. Powers, Mr. Connelly, &c., with whom I have had interesting talk.

I have left myself no space to write about politics. The enthusiasm for war in this country is wonderful; it seems to be a real genuine enthusiasm, a true crusading spirit; and I can wish the Italians success (if there must be war), though I can form no such wish as to their friends (? false friends) the Prussians. The question is so strange and so complicated a one that I hardly venture to form even a guess as to the probable results ; I can only see that the Italians are risking everything—risking the loss of all that they have gained in the last seven years. All that I am certain of is, that this is the worst time we could have chosen for coming hither. This last telegraphic news of the invasion of the Danubian Principalities by Russia and Turkey, in conjunction, complicates still more the whole confusion.

Fanny, I am happy to say, seems flourishing.

Ever your very affectionate

CHARLES J. F. BUNBURY.

JOURNAL.

Saturday, May 26th.

Soft relaxing weather—went to the Uffizii—spent an hour there with great enjoyment, principally in looking at the beautiful collection of original drawings which are arranged and exhibited in the most satisfactory manner in three rooms at the end of the farthest corridor.

1866.   The weather having become very fine and warm, we took a drive in the afternoon, to the beautiful villa of Careggi, the celebrated residence of the Medici, and where both Cosmo and Lorenzo died. It is situated on the first rise of the hills to the north of Florence, west of the Fiesole hills and the Bologna road, just where the ground begins to slope up towards Monti Morello : it is now the property of a rich Englishman, Mr. Sloane, who has brought it into admirable order.   The house is handsome and stately ; they show the room in which Lorenzo de Medici died, and the well in the court-yard into which the unfortunate physician is said to have been thrown.   The garden beautiful—in the stately Italian terraced style, and rich in brilliant flowers and fine flowering trees and shrubs.   The views from it delightful ; Florence itself indeed not much seen, being mostly hidden by the intervening shoulder of a hill; but the views over the rich valley and opposite hills, gay with innumerable villas, and on the other side, looking up to Monti Morello with the varied play of light and shadow on its majestic slopes and hollows are quite charming.

Sunday, May 27th.

Beautiful day.   We had a very pleasant drive in the country up the valley of the Mugnone, and the ravine which divides the hill of Fiesole from the next hill to the west of it.   On the top of this latter hill, which is very steep, stands a small white tower, said to be the Uccellalojo mentioned by Dante.

Lovely weather. Went with Joanna to the Museum of Natural History. We saw Professor Targioni Tozzetti (nephew of the one with whom I made acquaintance in 1848), who took us through a part of the Zoological department (the insects and other invertebrata which are his especial province), and Sig. Ancona, who showed us the organic fossils (as much at least as we had time to see). Both were very courteous.

After dinner we had a charming drive to Poggio Imperiale, and thence round the hills to San Miniato al Monti.

A lovely evening, and the country in its fullest beauty.

To Poggio Imperiale, a long gradual ascent, beginning first outside the Porta Romana, through a continuous avenue of noble Cypresses and evergreen Oaks. The Cypresses of both kinds, the upright and the spreading-branched, superb trees, both in height and bulk—many of the Ilexes also very large and old, with picturesque, rugged, and cavernous trunks. The former palace of Poggio Imperiale now, a great school for young ladies.

Drive from thence to San Miniato—narrow and rather break-neck roads, but a succession of most delightful views of the country. The white wild roses in the hedges just coming into blossom.

We visited a villa (? Arcetry) or farm house, on the top of one of the hills where Galileo lived, and ascended to the top of a tower from whence he is said to have observed the stars. The

1866. view from it is extensive in all directions, and very fine.

San Miniato,—fine and interesting church, in a very commanding situation, on one of the heights which directly command the city from the south,— the front cased with white and dark green marble. Glorious view of Florence and the Val d'Arno. Bastion adjoining the church built by Michael Angelo, to defend the city, in 1529. Cemetery of San Miniato,—the ground so thickly paved with tomb slabs, that one can hardly avoid treading on them. Very steep descent from San Miniato to the gate.

Tuesday, May 29th.

Weather very sultry and oppressive. Went with Fanny first to the Church of the Carmine (the Carmelites)—saw the noble frescoes of Masaccio, Masolino, and Filippino Lippi. Saw them very imperfectly, however, for the light is excessively bad, so that one can really judge of them much better from the prints. Then to the convent of San Marco : here we saw, first, in the chapter house, Fra Angelico's great fresco of the Crucifixion, with a multitude of saints on each side—a fine specimen of the master.—Afterwards I was taken through the convent itself (Fanny not admitted) and saw in the corridors and cells several other excellent works of the same Angelico—in particular, a coronation of the Virgin, and a Resurrection which struck me as more beautiful than any other of his works that

I have seen—the perfection of his peculiar style. I 1866.
was shewn also the cell which had been inhabited
by Savonarola.

———

Wednesday, May 30th.

The twenty-second anniversary of our happy
marriage,—thank God.

———

Tuesday, June 19th.

From Pistoria to Bologna over the main Apennines, a very extraordinary line of railway—
marvellous turnings and windings : passed through
not less than 44 tunnels, some of them very long.
Beautiful glimpses of scenery betweeen the tunnels :
a glorious view of Pistoria and the valley of the Arno
to the hills about Florence. The mountains high
and massy and excessively steep, but hardly anywhere
actually precipitous, almost everywhere clothed
either with trees or brushwood of Chesnut. Many
pretty and picturesque glens crossed by the railway,
some of them rocky, and the rocks beautifully
feathered with shrubs. Abundance of lovely wild
flowers seen in passing glimpses. We descended
the course of the Reno, which at present has not
much water.

Nearer to Bologna, the outer and lower ranges of
hills, probably of later formation and softer
materials, are in many places very abrupt and bare,
stripped and torn by water.

———

June 20th, Wednesday.

Bologna an excessively noisy town. Arcaded

1866. streets, giving a peculiar character. Many fine old palaces, especially that of the Bevilacqua family which is huge and sombre, somewhat in the style of the Strozzi and other old palaces at Florence, but with smaller stones. The great Piazza very noble and picturesque, retaining its mediæval character in great perfection. The two great towers:—the Assinelli, really very imposing (though not at all beautiful) from its enormous height, 320 feet, which produces its full effect. There being nothing to impede the view or weaken the impression. The garisenda looks quite unfinished and is as ugly as any steam engine chimney, only remarkable for its strong leaning.

Church of San Petronio :—front grand, though unfinished, fine bas-reliefs of the principal door. The interior, though vast, not to me so imposing as that of the Cathedral at Florence. Meridian line marked on the floor.

Gallery of the Academy:—rich above all in containing Raphael's lovely Saint Cecilia, very rich also in pictures of the Caracci, Guido and Dominichino.

June 21st, Thursday.

From Bologna, by Parma and Piacenza to Milan, a railway journey of about eight hours. All the way over an unbroken level plain, everywhere cultivated, and to the south of the Po, beautifully rich and luxuriant, especially between Bologna and Parma. No olives, but abundance of vines, trained in festoons between the trees, as beautifully as on the other side of the Apennines. Extensive cultivation of maize (now in flower), and of hemp. Much wheat cut, and the harvest going on busily.

To our left the plain bounded, at no very great distance by pretty ranges of hills, the outer ranges of the Apennines. To our right the plain unbroken to the horizon.

We cross the broad beds of several rivers or torrents, some quite dry, others with a scanty stream—they looked strangely white in the glare of the sun.

At Piacenza, a great many earthworks, apparently quite newly thrown up, to guard the approaches.

We cross the Po by a fine iron bridge, a little way from the bridge of boats. From the Po to Lodi, and thence to Milan, the country less beautiful;—though still uniformly cultivated it appears less rich, and has decidedly a less southern character, less distinctively Italian. Much cultivation of rice conspicuous by its light bright yellow green colour—grown in fields which are either now flooded, or have been so, the plant at present of small height and not in flower. Maize in these parts less forward than to south of the Po and of less luxuriant growth. Wheat cutting. Little or no hemp. Few trees except Poplars, and those much mutilated and disfigured. Vines grown low, not festooned.

———

Milan, Friday 22nd June.

We went to see the " Last Supper " of Leonardo da Vinci—rather less defaced than I expected to see

1866. it, yet very sadly so, and what is most provoking, the injuries appear to have fallen heaviest on the most precious parts of the picture, the heads of our Saviour and of John, but we can still see how glorious they must have been. Brera Gallery—a very extensive and very noble collection of pictures, rich in the works of the Lombard School, and also, though a less degree, in Venetian pictures.

Raphael's Sposalizio very charming in his early style.

Guercino's Abraham and Hagar—the expression of Hagar's face very beautiful.

Guido—Paul remonstating with Peter—noble.

Leonardo: a sketch of a head of Christ, not exactly the same as in his Last Supper, but almost equally beautiful.

Luini : many noble frescoes.

Sassoferrato : a Virgin and Child—most lovely.

Lorenzo Lotto : three portraits, very beautiful, especially one of a lady.

Bonifazio: " The finding of Moses "—a most singular treatment of the subject—a multitude of gay groups of richly dressed Venetian ladies and gentlemen, lounging, flirting, and amusing themselves in a wooded landscape—a Venetian fête champêtre, one might have supposed. But it is beautifully painted, and many of the women very lovely.

Albani : a Dance of Cupids—charming.

Gentile Bellini : " S. Mark preaching at Alexandria "—a most remarkable picture of immense size, with a vast number of figures, in Turkish dresses ;

both the costumes and buildings having all the 1866. appearance of true studies on the spot.

Paolo Veronese : several fine works.

Gaudenzio Ferrari : Martyrdom of St. Catherine, a striking picture.

In the afternoon we went to St. Ambrozio—a very curious and remarkable church, built in the 9th century. An atrium of considerable extent, in form an oblong square, in front of it ; curious details of sculpture on the principal doorway.

The high Altar (shown to us with great ceremony by the priests) is a Lombard work of the 9th century, exceedingly curious ; very rich and splendid, the front cased with gold, the sides with silver, most elaborately chased, with a multitude of figures representing Scripture subjects, and the life of St. Ambrose, and studded with a vast number of gems (said to be real, but this I cannot answer for—they really looked very beautiful) like all other antique gems, they are not cut in facets, but with a smooth convex surface. The body of St. Ambrose buried under the Altar.

In the tribune there are striking mosaics on a gold ground of the 9th century in fine preservation. In the folding doors of the principle entrance, two small panels are said to have been part of the very doors which St. Ambrose closed against the Emporor Theodosius.

At the east end, behind the Altar, there is a curious marble chair or throne, on which the Archbishops of Milan sat,—and it is said also St. Ambrose.

We visited the Ambrosian Library, but it would require many visits to appreciate it thoroughly. Most interesting autographs exhibited in glass cases:—Ariosto, a beautiful clear hand: Tasso, very indistinct : Galileo, an interesting letter to Cardinal Federigo Borromeo : Fra Paolo Sarpi: Carlo Borromeo, and the equally admirable Federigo Borromeo. Letters of Lucrezia Borgia to Cardinal Bembo, with a lock of her flaxen hair.

A very fine manuscript of Virgil:—it belonged to Petrarca, and has his marginal notes.

A small book containing the Life of our Lord from the Gospels and apocryphal Gospels, with a multitude of the most exquisite outline illustrations. Many religious books most beautifully illuminated, particularly one which belonged to Bianca Sforza.

A Latin translation of Josephus, a manuscript on papyrus, probably of the 5th century, one of the oldest books bound in the modern form, not in a roll.

A large volume of Leonardo da Vinci's original sketches and designs chiefly elaborate drawings of machinery and fortifications. A volume of architectural designs of Bramante. A noble fresco by Luino:—Our Saviour Crowned with Thorns—in excellent preservation; a very fine thing indeed, the head of Christ most touching in its expression, those of the executioners very powerful; in the foreground are introduced rather incongruously, portraits of a number of members of the religious confraternity of the Sacra Corona for whom the picture was painted.

Upstairs an extensive collection of pictures, drawings 1866. and engravings, particularly original drawings of Leonardo and Luino, most interesting to study, but we had not time. Raphael's original cartoon of the school of Athens, differing in some points from the fresco.

We drove to the Arco del Sempione. Saw the Arena, a modern amphitheatre.

June 24th Sunday.

A most oppressively hot day: and as it was Sunday and no galleries or shops open, we merely went after reading prayers into the gloriously beautiful Cathedral and took a general view of it. It is a great pity that the roof is merely painted in imitation of carving: such a sham harmonizes ill with so much that is really beautiful.

Milan is a fine city, but with no especially distinctive character of its own, like Florence, Genoa or Bologna.

I remarked the pavement of the streets. The women of Milan have fair complexions, and very many of them pretty with their very pretty head-dresses and black veils like the Spanish Mantilla.

Monday, June 25th.

We pass over the battle field of Magenta. Between the bridge over the canal (the Naviglio Gronde) and the river Ticino where the chief action took place, the railway runs along a high causeway, the grounds on each side low and wet, chiefly rice

1866. grounds and water meadows, much intersected by ditches; east of the Naviglio the country is higher, on a level, or nearly so, with the railway; highly cultivated and exceedingly intersected by rows of vines and mulberries. The Ticino, a fine, broad, full, strong flowing stream, very different from the inconstant torrents flowing into the Po from the south.

We embarked (on the lake) at Arona: a furious storm of rain came on, and continued till we arrived at Stresa, so we saw nothing of the lower branch of the lake. The colossal statue of San Carlo Borromeo on a commanding eminence above the lake, though dimly seen through the mist and rain, had a majestic effect.

Tuesday, June 26th. Stresa.

This is one of the most delightful situations imaginable. The hotel stands on the margin of the lake, on its western shore, just at the angle where the branch containing the Borromean island turns off from the main lake: these islands very near; the bright little town of Pallanza directly opposite, on a beautiful promontory which juts into the lake from the mountains bounding the Borromean branch on the north; Laveno, another little town on the opposite shore of the main lake at the foot of a great mountain. Views up and down the lake in three directions.

The variety of the forms, colours, and grouping of the mountains, the beauty of the lights and shades on their sides, and of the varying effects of clouds, the brightness of the little towns, the picturesqueness of the little white churches and cottages perched at various heights in the Chesnut woods: the way in which the effect of everything is heightened by the exquisite glassy lake which seems to harmonize the whole; all is quite beyond my power to describe. The union of grandeur and softness which is so especial a characteristic of this scenery, is admirably well-represented in the "Song of Elena."*

The bold and many-peaked mountain immediately and above Laveno is particularly beautiful.

Rambled out to botanize, but was soon caught in a shower heavy enough for Madeira, and got a good wetting.

After an early dinner, we went in a boat on the lake and visited the Isola Bella and Isola Madre.

The general effect of Isola Bella is questionable, but the gardens are very beautiful; not only the oranges, lemons, citrons, and oleanders grow on the sunny terraces with a luxuriant beauty which reminds us of the Genoese Riviera, but Bay Laurels, Cypresses, Pines, Ilexes, and Cork trees grow to a surprising size, most surprising considering the small depth of soil in which they grow on the terraces. Magnolias also very fine. We saw the famous Bay tree on which Napoleon carved some letters, a tree of wonderful bulk and height indeed.

Isola Madre, less pompous and artificial in its general effect, and the gardens equally fine, or

* See "Philip in Arlech," by Henry Taylor.

1866. perhaps still finer—at any rate stored with a greater
variety of semi-tropical plants.*

On both the islands there are a vast number
of turtle doves, quite tame, but at liberty, flying
about and perching on all the trees, close to one,
without the least fear. Their incessant cooings
make a surprising noise. They are the fawn-
coloured black-ringed African turtle-dove. On Isola
Madre, the rocks near the water are covered with
huge tufts of the great Agave, growing as if wild.

The waters of the lake are beautifully clear, dark
green in the deeper parts (when one looks directly
down on them), pale, transparent green near the
shore, not blue-green like the Lake of Geneva.

———

Wednesday, June 27th. Stresa.

Fanny and I took a charming little ramble up
a hill near the hotel by winding paths among the
vineyards, returning down the dry bed of a torrent.
The details of the scenery are as lovely as the general
effect. Vines beautifully trained on trellises, maize,
wheat, barley, hemp, kidney beans and mulberries
are in rich profusion on the steep terraced slopes ;
and higher up the mountains are clothed with
chestnut trees. We found Asplenium septentrio-
nale and Ruta muraria, and gathered specimens
of the great Phytolacca, which is common here.

The people are cheerful and pleasing in appearance and courteous in manner.

In the afternoon, we went in a boat to Baveno

* Some very fine Conifers.

a few miles further up the western branch of the 1866.
Lake ; landed and walked up to the granite
quarries. (We landed indeed some way beyond
Baveno, that village being south of the quarries).

The way up is through a lane like a Devonshire
lane, fringed with Ferns—all familiar English kinds.
Observed Osmunda Regalis in one place. The
quarries very extensive ; the whole side of a great
hill scarped by them.

Thursday, June 28th

We left that most charming place Stresa and
set out for the Simplon with two carriages, each
with two horses. We arrived at 8 p.m. at Isella,
where we slept ; bad horses and other things made
the day's journey tedious and disagreeable ; the
good horses we had engaged were taken for the war,
and bad horses sent in their place.

The character of the villages and little towns
through which we pass, thoroughly and strikingly
Italian. The beautiful trellised vines the form and
construction of the houses, the churches, the look of
the people, all Italian.

Domo d'Ossola, as distinctly Italian as any. The
valley up the Isella a grand and gloomy ravine.

June, 29th Friday.

From Isella over the Simplon to Brieg in the
Valais. From the village of Simplon to the top of
the pass, a gradual ascent of about three hours
through fine Alpine pastures, interspersed with

1866. rocks : astonishing crags and peaks towering above, but no precipices near the road. The pastures enamelled with a vast variety of most beautiful wild flowers. In particular, the yellow alpine Anemone, Anemone sulphurea, in great plenty. Primula farinosa very abundant and beautiful in boggy places ; another very beautiful Primula (? viscosa) in the moist chinks of rocks.

————

Saturday, June 30th.

The Rhone, a most impetuous and furious torrent, with turbid waters of an ashy grey colour.

————

Monday, 2nd July.

From Brieg to Sion, halting at a very nice inn at Susten to rest the horses

At Visp, a striking view up the lateral valley descending from Zermatt, a glimpse of Monte Rosa amidst the clouds. Opposite to Susten, up a lateral valley to the north, a grand mountain appears utterly bare and precipitous, nothing but rocks and snow. Prodigious exuberance of insect life in the meadows behind the inn at Susten. Butterflies of many kinds, locusts and grasshoppers in surprising numbers. A marble-brown and white butterfly, especially abundant : also Fritillaries, a lemon-yellow Colias, and a pearl-grey Polyommatus.

Between Sierre and Sion, some very remarkable high mounds, looking almost like artificial embankments, stretching partly across the valley, probably relics of ancient moraines. Near Sion, many vineyards on the lower slopes of the mountains. The vines trained low in the French manner. I have since been told that they yield pretty good wine. From Sion to Vevey by railway.

———

Tuesday, July 3rd.

From Vevey to Geneva by railway, the weather being too rough for the Lake.

———

Wednesday, July 4th.

Geneva. *Hotel de l'Ecu*—good, and fine situation. We drove out and visited Mr. Alexander Prevost and M. de Candolle—the latter just returned from England. We dined with Mr. Prevost and met Mrs. Plantamour and her Son-in-law.

———

Thursday, July 5th.

We visited the cemetery and saw dear Sir George Napier's grave. I called by appointment on M. de Candolle, and had some botanical talk with him—he showed me something of his herbarium and library.

Visits from M. de la Rive and Mr. Prevost. We drank tea with M. and Mme. de Candolle.

———

Friday, July 6th.

We left Geneva very early by the train for Macon.

Fine picturesque gorge of the Rhone on the French frontier near Bellegarde. Scenery like parts

1866. of Derbyshire. The French custom house is at Bellegarde. We dined at Macon, and went on by a late train to Dijon.

———

From Dijon by train to Fontainbleau, where we dined and slept.

———

Fontainbleau. We walked through part of the gardens, admired the exterior of the palace, and fed the carps. We went on to Paris.

———

Paris. We went to Leroy's and bought watches. Fanny spent all the afternoon shopping. I strolling.

———

From Paris to Boulogne. A sea-fog deterred us from crossing.

———

We crossed over from Boulogne to Folkestone, an excellent passage. Mr. Hawker and Mr. Symonds on board. Went on from Folkestone by the express train. Arrived at 48, Eaton Place, before six ; all safe and well, thank God. Parted with dear Kate, who went to her home.

———

London. Excessively hot.

Called on Lady Napier. Dear Katharine and Leonora came to luncheon with us ; very pleasant. Talk with Edward. Visited the Royal Academy Exhibition

Visit from dear Minnie and Sarah ; Rosamond, Annie and Dora came to luncheon with us.

Met Louis Mallet at the Athenæum. Talk with him on foreign politics.

Still very hot. William and Emily Napier and three of their children came to luncheon with us.

Clement arrived. We went to see the Henry Lyells and Leonora. Edward dined with us and stayed late.

We went to St. Peter's, Eaton square, and heard an excellent sermon from Mr. Fisher.

Minnie and Sarah came to luncheon with us.

Visited the British Institution. Visit from the MacMurdos. Minnie, Sarah and Edward dined with us.

Revisited the Royal Academy. Visit from Norah Bruce and Catty. We dined with the Bishop of London (Tait) at Fulham — a large party, but pleasant enough.

My Dear Wife's 52nd birthday.

We went to the Woking cemetery and saw the beautiful monument of Mr. Horner.

1866.   The MacMurdos and Douglas Galtons dined with us,—very pleasant.

Thursday, 19th July.

We went to the British Institution. Saw also a picture of the Rocky mountains exhibited at McLean's.

Minnie and Sarah dined with us.

Saturday, 21st July.

We went to see the Loan Portrait Gallery, South Kensington, and spent a good time there. A most interesting collection. Visited Lady Napier.

Sunday, 22nd July.

We went to St. James's chapel and heard an excellent sermon from Mr. Brooke*. Minnie dined with us.

Tuesday, 24th July.

We drove down to the William Napiers at Putney, to the christening of their baby. Met Kingsley, the MacMurdos, Norah and Catty. Kingsley performed the ceremony. I had a very pleasant walk by the riverside with William and Kingsley. On our way back we called on Mrs. Tait at Fulham.

Thursday, 26th July.

Called on Sir John Boileau and had a pleasant talk with him.

* Rev. Stopford Brooke.

We went to Twickenham and dined with the Richard Napiers; found them in pretty good health; and as usual most kind and pleasant. He is within a few days of 80 and she 85. Mrs. Grey was with them.

Saturday 28th July.

Emily Napier and Cecil came to luncheon with us.

We left London by the five o'clock train and arrived at our dear home at Barton soon after nine, all safe and well, thank God, truly happy to be at home again.

Sunday 29th July.

A very wet day, but I managed to have a look at the arboretum and the Fern house. Read prayers with Fanny, and in the evening read one of Kingsley's sermons to the servants.

LETTER.

Barton,
July 31st, 1866.

My Dear Joanna,

I thank you very much for your very kind and agreeable letter from San Marcello. I am heartily glad to hear that you and Susan are both well, and that you find San Marcello so pleasant. It must indeed be a charming retreat and refuge, both from

1866. the heat of the Italian plains and from the worse
heat of political and military excitement. I thank
you very much for your kind expressions towards
myself, and I can assure you that I look back with
very great pleasure on our stay at Florence,
notwithstanding the cold which spoilt part of it for
me. It is very pleasant to look back on the time we
spent with you and Susan, our daily meetings in
the Villino della Torre, or at No. 126; our visits
together to galleries and churches, our delightful
drives in the country, and our cheerful suppers.
Those five or six weeks have stored my memory
with a multitude of agreeable images. Even the
bells of S. Rumido *at this distance*, are not unpleasant. I am more happy than I can well express
to find ourselves at home again, and most thankful
to find all well and in good condition. I was so
impatient all the time we were in London, to get
back to my *real* home, that I could not half enjoy
even the company of the friends we did see : though
I was very glad to have a glimpse of dear
Katharine and Leonora and their children, as well as
to see Minnie and Sarah and the William Napiers
and MacMurdos. It was a special pleasure to have
a conversation with Kingsley, whom (as Fanny will
have written you word) we met at the christening of
William Napier's baby.

We have not been favoured by the weather since
we arrived here : it has been, and is so cold and
wet that we have returned to winter clothing and
fires, but yet we have managed to see the gardens
and stables and part of the grounds. Everything is

in perfect order. The growth and luxuriance of foliage of the trees and shrubs are remarkable ; we are almost too late for the roses, which are said to have been very fine : but my favourite plants, the Ferns, are very flourishing. The hay crop this year has been extraordinary.

The old pensioner horses in the park are flourishing in a green old age, and seem to have no thoughts of departing this life. Fanny's pensioner birds too are as lively and numerous and impudent as ever. We have not yet had any visitors except the Suttons, but Fanny went yesterday (August 1st) to see Mrs. Rickards, and found her well and comfortable.

The political horizon looks very stormy and unsettled, but I hope there is a probability of a speedy termination to that horrible war. It is shocking to think of the thousands of families that have been rendered desolate and miserable to gratify the ambition of the King of Prussia and his Minister. No doubt as God's Providence brings good out of evil, it is very possible and probable that the ultimate result of the war may be good :—that the complete domination of Prussia over Germany may be better for mankind than the divided state that has hitherto prevailed. One must believe that Bismarcks and other things "are but as slavish officers of vengeance" to the supreme good ; otherwise one could hardly reconcile one's self to the course of history ; but this does not prove that a war of aggression and conquest is not criminal. I wish the calamities of

1866.

1866. war fell only on the kings and ministers who cause
them. As for the Italians, who really had some reason
for attacking Austria, I am afraid they have been too
much under the belief that courage and enthusiasm
are the only requisites for victory. I hope, or rather
wish, that after having obtained Venetia, they
may become less ardent for war, and turn their
attention more to commerce and internal improvement. But I am rather afraid the effect of this war
will be to make the Sovereigns of Europe devote
themselves more than ever to military pursuits, as
they have seen how quickly a great power may be
crushed by the superior military organization of an
ambitious neighbour.

(August 4th). We have been over to Mildenhall
for a day; I wanted to attend a meeting of the
Board of Guardians, held to discuss the precautionary measures advisable against a possible
visitation of cholera. We found the place looking
well and in good order: and Mrs. Bucke flourishing,
very lively and in good health and excellent spirits
at 83. The weather continues chilly, raw and
stormy: it does not agree with Fanny or me, and
very unfavourable for the harvest, which is just
beginning. The crops promise well, if only we can
have fair weather for gathering them in.

The general health both about Mildenhall and in
the neighbourhood is as yet satisfactory. But the
accounts of the cholera in the east of London are
truly awful: it seems to be more malignant and
more terrible than in any of the previous visits to
England. I am very sorry that we are not to see

you in England this autumn. but I hope you will 1866. very much enjoy Rome and Naples, and that you will take care to keep out of the way both of brigands and of cholera.

With much love to dear Susan, believe me ever,

Your very affectionate Brother,

CHARLES J. F. BUNBURY.

JOURNAL.

Wednesday, 1st August.

Weather improving. We went to the cemetery,* and were much pleased with the alteration in it.

Thursday, 2nd August.

Extremely wet. Visit from Lady Cullum. We went over to Mildenhall.

Friday, 3rd August.

I attended a special meeting of the Board of Guardians to consider precautionary measures to be taken against cholera.

We returned home.

Monday, 6th August.

Very bad weather. Mrs. Abraham came to luncheon with us—charming.

Tuesday, August 7th.

Weather rather better. Began to write notes of my Father's life.

* We had lowered the wall round the family cemetery, and planted flowers in it.

The library catalogue continued under the super-
intendence of both of us, and resumed catalogue of
my herbarium.

Thursday, August 9th.

Finished vol. ii. of Napoleon's Jules César. This
volume is much superior to the first. The narrative
of Cæsar's campaigns in Gaul, which occupies the
greatest part of it, is very careful, clear, and
interesting ; what especially distinguishes it, and
renders it more satisfactory than most histories
of ancient times, is the abundance of precise topo-
graphical details, illustrating the military operations.

The Emperor has evidently taken great pains with
this portion of the history, and employed many
persons to survey and examine accurately all the districts and localities which were the scenes of action
in the course of the war ; so that he has determined
the sites with precision and has given much more
copious and satisfactory local details than any other
historian I have read.   This is a valuable contribution to history.

The narrative of the military operations is also
very clear and pleasant in style.   The only fault
I should find with this part of the book is that the
writer shows a want of moral sensibility : he mani-
fests little sympathy with the heroic struggle of the
Gauls for their independence, and no indignation
against the cruelty of the Romans.   Later on in the
volume, indeed, he introduces some reflections on

the conquest of Gaul, which show somewhat of 1866.
a better spirit. No doubt he is right in his conclusion that, on the whole the ultimate results of the
Roman conquests were beneficial to the Gauls and
to mankind ; God's providence brings good out
of evil. But it is not therefore allowable to palliate
or to gloss over the atrocities of the conquerors and
the misery which they produced.

In the latter part of the volume the events which
led to the civil war are related with great clearness ;
of course in a spirit most favourable to Cæsar, yet
on the whole fairly.

I have long thought that Arnold, in his very
severe condemnation of Cæsar, did not sufficiently
consider the circumstances in which he was placed,
and was mistaken in believing that the Republic
could have been preserved in more than name. It
was merely a question, as it seems to me, whether
Rome should become subject to Julius Cæsar or
to some worse man.

We continued the library catalogue ; went to call
on the Arthur Herveys—out. Then on Sir James
Simpson, who is laid up with gout—saw him—then
on Lady Cullum.

_____

Friday, August 10th.

Began to read Baker's "Great Basin of the
Nile." Madame de Tourgueneff and her daughter
arrived.

_____

Saturday, 11th August.

We walked round the grounds with Madame and
Mdle. de Tourgueneff, showed them the gardens and

1866. stables. I visited Mrs. Rickards. Lady Cullum dined with us—very pleasant.

Monday, 13th August.

We went with Madame and Mdle. de Tourgueneff to Hardwick, and spent the afternoon very pleasantly there with Lady Cullum, saw the hot houses and the grounds.

Tuesday, 14th.

Read to the ladies in the evening, my Father's account of his interview with Napoleon.

Wednesday 15th.

The Arthur Herveys dined with us.

Thursday, 16th.

Madame and Mdle. de Tourgueneff went away early.

Saturday 18th.

A beautiful day. We went to Framlingham to see the Albert College. Arthur Hervey and his daughter Sarah met us at Stowmarket, and went with us to Framlingham. — A delightful drive. There Hervey and I attended a meeting at the College, while the ladies sketched. We returned to Barton together.

Sunday 19th.

A beautiful summer day. We went to morning

church. I spent most part of the day in strolling 1866. about the grounds. Read to the servants in the evening one of Kingsley's Town and Country Sermons.

Monday, 20th.

We walked about the grounds for a long time, and marked many trees.

Tuesday, 21st.

Leonora and her dear little girls arrived.

Thursday 23rd.

A beautiful day. Walked over my farm. Wheat stacked in good order. Peas and Oats stacked. Men busy getting up the Barley.

Friday, 24th.

Went to Bury to a meeting of the Thingoe Board of Guardians, concerning the New Health Act. Pertz arrived. Visit from the James Bevans.

Saturday 25th.

Fanny and Leonora went over to Mildenhall and returned to dinner. I took a walk with Pertz.

Monday 27th.

A very fine day. Finished reading Baker's "Nile." It is a very interesting book: a clear and satisfactory record, of a very extraordinary journey, performed by a very extraordinary man. It is indeed wonderful how any man, unsupported and

1866. unaided by any public authority or public force, could have overcome the enormous hardships, difficulties and dangers of such a journey ; and still more wonderful that his wife, a refined European lady, should have accompanied and helped and encouraged him through it all.

The difficulties, sufferings, and privations seem almost too great for human endurance.   The story is told with great clearness and animation, and in a gentlemanlike tone of feeling ; and altogether it is one of the best books of *adventurous* travel I have ever read.

Though the personal adventures form the principal and most striking part of the book, it is not wanting in curious information relating to physical geography.   It concurs with the other recent exploring journeys in showing that the ancient and popular idea of Africa, as a land of deserts, is true only in a very limited sense ; true only of a certain zone near the northern coast, and another near the southern extremity.

Central Africa appears to be a fine country of tropical luxuriance, but deadly in general to European constitutions.   Baker seems to have met with nothing like *desert* to the south of Khartoum. The vast and dreary expanse of level and flooded swamps covered with reeds, through which the White Nile winds, "tardis ingens ubi flexibus errat,"—from Gondokoro to Khartoum, is succeeded, after he leaves Gondokoro, by the fine mountain country of Latooka, which by his descriptions must be full of picturesque beauty.   From thence to the great

Lake he finds the country to have an average 1866. elevation of about 4000 feet above the sea level, with some mountains rising considerably higher; the country seems to be well covered either with wood or grass and abounding with large wild animals. He complains of the swamps in his way from Kamrasi's capital to the lake, and of the obstruction caused on his return journey by dense and gigantic grass jungle ; but nowhere anything like desert.

He is a keen sportsman, but no naturalist, and gives no definite information respecting either the plants or animals of the country, except the large game.

Whether or not we may as a matter of geographical theory, allow Baker's Albert Nyanza to be properly called the source of the Nile, it seems clear, at any rate, that this great lake is the principal basin or reservoir from which the White Nile derives its constant supply of water.

His observations, combined with those of Speke and Grant, seem to have clearly made out that the surplus waters of the "Victoria" lake flow by a river of short course into the "Albert," near the outlet of the latter, which thus in fact receives all the drainage of Equatorial Africa.

But very much remains yet to be ascertained as to the extent and form of both lakes, the rivers falling into them, &c.

————

Tuesday, August 28th.

Dear Katharine and her Husband and Children arrived.

————

Very wet. I returned to Gibbon (my reading of whom had been interrupted since Jannary 26th) and read part of chapter 24. Julian's residence at Antioch, and the beginning of his expedition against the Persians.

Katharine sang charmingly to us.

———

Helped Fanny in the library catalogue. Showed some of my dried plants to Katharine and Leonora.

William Napier with two of his children, and Kate MacMurdo arrived. We all walked round the grounds.

———

Went to Bury to a Magistrates' Meeting about Cattle Removal Licences. Tedious debates. Cecil and Clement arrived.

———

Fine day. Fanny and the Pertzes went over to Mildenhall, and returned to dinner. Finished 24th chapter of Gibbon.—Julian's brilliant but unfortunate expedition against the Persians is related in a very interesting style. I remark also in this chapter a good specimen of Gibbon's epigrammatic satire.—" They (the people in Antioch), contented " themselves with disobeying the moral precepts, " but they were scrupulously attached to the specu- " lative doctrines of their religion."

Emily Napier arrived.

Sunday, September 2nd.

Showed the arboretum to William and Emily.

Monday, September 3rd.

Received Humboldt's and Bonpland's " Nova Genera," a new purchase. A fine and complete copy ; Seven superb quartos.

Wednesday, September 5th.

Lady Arthur Hervey and her younger children came to us in the afternoon, a nice merry children's party.

Thursday, September 6th.

The children acted a charade very prettily. Sir Edmund Head arrived.

Friday, September 7th.

Walked with Pertz to the Shrub. Pertz gave me a curious account of a very rare and valuable MS., which he has lately procured for the library at Berlin. It is a MS. of the first book of Virgil's Georgics, and is one of the most ancient MSS. known. So ancient that he believes it may even have been of the time of Augustus.

It is on parchment, beautifully written entirely in large capital letters, and is for the most part in good preservation, as far as it goes ; it contains (I think he said) about 560 verses. No other MS. of Virgil of nearly equal antiquity is known, except one—also a fragment of the Georgics—in the Vatican; and when the Vatican MS. was compared with Pertz's new acquisition, it was found that they must have

1866. been originally parts of the same copy, the one supplying the gaps of the other. He obtained his precious MS. from Holland, at a sale by auction, and gave only five dollars for it. He has promised me a copy of the paper in which he has given its history.

Pertz thinks it probable that the ancients used *books* bound in somewhat the same shape as ours, as well as rolls; although the earliest MSS. in the book form *at present known* are only of the 5th century,

Showed some of my specimens to Katharine and afterwards walked with her in the arboretum. Lady Cullum and the Wilsons dined with us.

Saturday September 8th.

Sir Edmund Head went away. We planted trees for the Napiers and Kate MacMurdo.

The house being full of company, I have not read much lately. We received the sad news of Madame Pulsky's death.

Tuesday, September 11th.

The Lyells and Pertzes all went away. Very sorry to part with them.

Finished 26th chapter of Gibbon. The description of manner of life of the Tartars ; the passage of the Danube by the Goths : the battle of Hadrianople and death of the Emperor Valens are fine pieces of historical writing.

Wednesday, September 12th.

Mrs. Power and Mrs. Byrne, my Wife's Aunts

went away. Poor Mrs. Power, I fear, is in very bad
health. Mary and Theresa Boileau, Lady Head
and the Miss Heads arrived. The Arthur Herveys
and Lord Francis dined with us.

Thursday, September 13th.

A fine day. I showed the arboretum and garden
Ferns, &c. to Lady Head and the Miss Boileaus.
Mr. Harcourt Powell and Mr. Beckford Bevan
dined with us.

Friday, September 14th.

We all went to the rifle shooting in Ickworth
Park. Luncheon at the big house — delivery of
prizes—speeches.

Lady Cullum and Patrick Blake dined with us.
Very diverting, both of them.

Saturday, September 15th.

Emily Napier and her darling children and the
Miss Boileaus went away.

We went to the Church with Sir Edmund and
Lady Head—they went away after luncheon.

Sir Edmund Head tells me that Sir George
Cornewall Lewis really did say : —"Life would be
very tolerable if it were not for its amusements."
—And that this saying expressed his real opinion.
It is a very happy one, as applied to the generality
of what are called amusements. Sir Edmund Head
is a man of extraordinary range and variety of
information and accomplishments, and his conversation is very interesting. His manner on slight

1866. acquaintance is dry and cold, so that it was long before I "took" to him ; but when one knows him well, he is very communicative, cheerful, and even joyous. His knowledge both of books and men is remarkably extensive and various. There seems to be (with the exception of the natural sciences) hardly any subject of rational curiosity on which he is not well informed, but his more special study has been that of the fine arts (in the most extensive sense) ; in every branch of these he appears to be really learned. He is also, from his long official experience, well acquainted with practical politics and political men, and his political views appear to me moderate and reasonable.

———

Sunday September 16th.

Went to morning church with Fanny, Kate, William Napier, and Clement.

Read chapter 5 of S. John, in Greek, with Alford's notes.

———

Monday, September 17th.

A fine day, but very cold. William Napier left us. We went with him in the carriage as far as Stanton, on his way to Garboldisham—then parted, very sorry to part with him. We went to Langham. Saw Maitland Wilson. On our way back went to look at dear Mr. Rickards' grave.

———

Tuesday, September 18th.

Fanny, Kate MacMurdo, and I drove to Ickworth and drank tea with the Arthur Herveys.

Finished chapter 27th and vol. ii. of Gibbon. Theodosius doubtless obtained the title of *Great* for his devotion to the priests. Yet he seems to have been really a good Emperor—one of the few good ones after the Antonines, though disgraced by bigotry and a spirit of persecution ; which same persecuting bigotry was probably one of his chief merits in the eyes of the priests.

Read chapter 1st of book 9th of Trollope's " Florence." Charles VIII. stay in Florence, and the heroic reply of Piero Capponi.

LETTER

Barton,
September 20th, '66.

My dear Katharine,

I am very glad to hear that you have got another Ward's case, and that the young Ferns you got here are acceptable additions to your collection ; I trust they will thrive. Many thanks for the information about Tropæolum speciosum. That reminds me :—if you should happen to see Mr. Bentham, would you be so good as to ask him the name of the Tradescantia with rooting stems and prettily coloured leaves, which we cultivate here for ornamenting dishes ; I cannot find it in any book that I have by me. I hope you took a good piece of it with you, and that it will thrive in your Ward's cases ; if not, I will tell Allan to send you some more.

1866.   I am promised some fine Orchids. Mrs. Mac
Murdo has offered to send me, when she goes to
Ireland, those which were sent to her from Brazil;
and I hear they are fine plants. I do not feel sure
whether they will thrive in my Fern house, or
whether they will not require more heat; for though
several of the Ferns which succeed perfectly there,
are natives of completely tropical latitudes, and
though Ferns and Orchids flourish together in the
Brazilian forests, yet, in cultivation, the Ferns seem
to accommodate themselves to a more temperate
climate than suits the Orchids. At least, this is the
supposition on which cultivators have always proceeded, but it may be worth trying, and if the plants
will thrive with less heat so much the better.

Fanny and I are well, and have plenty of agreeable occupation; but *(September* 21st), we are much
grieved now by the very serious illness of our excellent old coachman. Poor dear old man, I am
very much afraid that we shall never see him about
again; the doctors agree that he has a mortal complaint, from which he can never possibly recover,
though he *may* rally for a time; but I believe he now
thinks himself that his end is near. He has been
an excellent man, and has lived here with us so
long, I can hardly reconcile myself to the idea of
Barton without him. I so identify him with the
place where he has been ever since I first came
to live here with my father and mother, more than
42 years ago.

I often miss the dear children—the merry little
party who made this house so joyous a little while

ago; I feel that I should like to see and hear them 1866 again, racing about the house. It was a great pleasure to have them all here, and your visit was a great pleasure to me, dear Katharine; it all seemed to be too soon over. We have not been alone however, for Kate MacMurdo and Clement have remained with us; and since I began this letter, the Bowyers have arrived. I think you have met them here; very pleasant people they are; he remarkably clever and agreeable. Next week we expect again a housefull.

Pray give my love to Charles and Mary, as well as to your husband and the dear children.

And believe me ever,

Your very affectionate brother,

CHARLES J. F. BUNBURY.

JOURNAL.

Friday, September 21st.

I visited poor old Wallis, who is hopelessly ill.

Read a chapter of Trollope's "Florence." Savonarola's ascendancy, and the constitutional government established by him. Trollope's political opinions appear to me very sound.

Monday, September 24th.

My Barton rent audit day: very satisfactory. Mr. and Mrs. Hutchings, Mr. and Mrs. Mills, Sir Edmund and Lady Head arrived.

1866.   Tuesday, September 25th.

I showed the arboretum and gardens to Mr. Mills and the Hutchings. Sir Edmund and Lady Head went away. The Frederick Freemans arrived. Several Portugal Laurels were cut down in the shrubbery, and two American Oaks in the arboretum. Lady Cullum and Mr. Beckford Bevan dined with us.

Wednesday, September 26th.

The Arthur Herveys dined with us: also Captain and Mrs. and Miss Horton from Livermere.

Thursday, September 27th.

Our dear good old Walter Wallis died. Mr. and Mrs. Hardcastle (she a daughter of Sir John Herschel) dined with us.

Friday, September 28th.

The Millses went away. Resumed my biographical notes about my Father. Walked with Augusta Freeman.

Saturday, September 29th.

A beautiful day and very warm. Fanny and all the party, except Clement and I, went over to Mildenhall and returned to dinner.

Sunday. September 30th.

Splendid summer day, very hot. We all went to morning church, a large party. Strolled about the grounds all the afternoon.

Tuesday, October 2nd.

We attended the funeral of our dear old Walter
Wallis. He was buried beside his daughter, near
our family burial place; the attendance was very
good, and the service well performed.

Wednesday October 3rd.

Mr. and Mrs. Hutchings went away; they are
both very pleasant.

Thursday, October 4th.

We dined with Lady Cullum. Met nobody but
the Millses.

Friday, October 5th.

I have hitherto omitted to mention, among my
employments, the writing a sketch of my Father's
life as an introduction to the MS. fragments
he left, and which we intend to print. These
biographical notes I began on the 7th Aug: dis-
continued them for some time during the crowd of
company: then resumed them on the 28th September, and have gone on steadily and gradually
with them. Whether anything will come of them
—*quien sabe?* I am also going on with the catalogue
of my herbarium, and am now in the tribe Aveneæ
of the Grasses.

One of the Magnolias in the arboretum flowered.

Saturday, October 6th.

A beautiful day. We went to choose places for

1866. planting new trees, and inspected those planted in the Vicarage Grove.

I walked to the Copses on Loft's farm ; found nothing.

———

We went to afternoon Church. A very good Sermon from Mr. Marshall, in which he spoke with much feeling and good taste about our dear Walter Wallis.

Finished General Shaw Kennedy's " Notes on the Battle of Waterloo." The book is not particularly well written, but is interesting, as containing the criticism of a very able, thoughtful and experienced officer, who was himself present as a staff-officer at the great battle. In his narrative I do not perceive much that is new, except that he distinguishes, perhaps more clearly than other writers have done, the five grand attacks made by the French—the five acts (as he calls them) of the battle, namely :—First, the principal attack of Hougoumont. 2.—The attack on Picton's Division and the left of the Allied line. 3.—The grand charges of the French cavalry 4.—The attack on La Haye Sainte, and from thence on the Allied centre. 5.— The final charge of the Imperial Guard. I am rather surprised that he does not mention Charra's excellent book on the Waterloo campaign ; He does not seem to have seen it ; but several of his criticisms are essentially the same as those of Charras. He finds great fault with Wellington and Blucher for occupying, at the opening of the campaign, a line much too extensive, which made it difficult for them to concentrate, and exposed them to the risk of being beaten in detail. He also blames the Duke (as Charras does) for keeping a considerable force detached at Hal, too far off to take any part in the battle. But above all, what he especially notes as an error in the Duke's arrangements, is the neglect of La Haye Sainte; the not occupying it with a sufficient force, and not taking adequate precautions for its effectual defence, in fact the not perceiving its great importance to the Allied position. He points out that the most dangerous crisis in the whole action was when the French had got possession of La Haye Sainte, and under cover of it were able to move up their infantry and guns to within sixty yards of the centre of the Allied line.

On the other hand he severely criticises some parts of Napoleon's operations, especially the slackness and dilatoriness of his movements on the 17th, whereby the Allies were enabled to retreat from the position of Quatre Bras to that of Waterloo, without being seriously engaged.

On the whole he concludes, as Charras does, that at the time of that campaign, Napoleon had lost much of his former activity and energy; and this agrees with a remark made by Soult when in England, to Sir William Napier. General Shaw Kennedy expresses very high admiration of the manner in which Napoleon brought his army into position, and began the action at Waterloo; but points out various errors which he appears to have committed in the details of the battle from want of

1866. a sufficiently close personal inspection of the movements : above all that he did not profit by the important advantages gained by his troops at La Haye Sainte.

———

Monday, October 8th.

Finished the 14th chapter—2nd of the 9th vol. of Froude's History.   It is very remarkable (as clearly shown by the contemporary correspondence and documents quoted by Froude), what desperate efforts Elizabeth made to obtain the release of Mary Stuart from Loch Leven, and even her restoration to the throne.   She was not content with remonstrating, reproaching and even threatening the confederate Lords, but she seems seriously to have intended to interfere by force, and to have been withheld from so doing only by the certainty, which was pointed out to her that the first step taken in that direction would have caused the immediate execution of Mary.   It does not seem unjust however, to conclude that her hatred for anything like rebellion, and her horror of a precedent of a deposition of a soverign by the people, had at least as much to do with the anxiety as any personal regard for her rival.   It seems certain, that but for the influence of Murray, Mary Stuart would have been put to death by her own subjects.

Took a walk with Augusta Freeman to East Barton, and Barton mere.   Dear Minnie and Sarah arrived.

———

Tuesday, 9th October.

Took a long walk with Augusta.   Cecil and young Kinloch arrived.

Augusta and Frederick Freeman went away.

I finished Scott's most excellent report on the Mildenhall cottages.

Many guests arrived. Fanny went to Mr. Gilstrap's ball, with all the party except me.

Thursday, 11th October.

Visits from Sir George and Lady Nugent, the Miss Waddingtons, Mrs. Dashwood and Mr. and Mrs. Corrance, occupying all the afternoon.

Walked round the arboretum with Mr. Thornhill, Major Tyrell, Mr. Barnardiston and Mr. Corrance.

Friday, 12th October.

Lord John Hervey and three others from Ickworth dined with us. Fanny with most of the party went to the Bury ball.

Saturday. 13th October.

The Barnardistons, Thornhills, and Major Tyrell went away in the morning. Cecil and Clement and Lady Head and her daughters went away in the afternoon. Dear Minnie and Sarah remained. We visited poor General Simpson, who is quite crippled.

Tuesday, 16th October.

We went, Fanny, Minnie, Sarah and I to Cambridge. Put up at the University Arms. Fanny shopped for Clement, while Minnie, Sarah, and I went to see Sedgwick, and he took us through

1866 the College walks and showed us the new buildings of St. John's, etc.

We all drank tea with Mr. Clark,* and dined with Dr. Phelps (at Sidney Sussex College).

————

Wednesday, 17th October.

Fanny worked all day, helping Clement to furnish his rooms. Sarah and I went with Sedgwick to see Trinity College Library and the Woodwardian Museum. We had a very pleasant luncheon with Sedgwick.

————

Thursday, 18th October.

We returned home. The Phillip Mileses and Edward arrived.

————

Friday, 19th October

Wretched damp foggy weather—but not cold. Began to arrange the plants collected on our tour.

————

Sunday, 21st. October.

A beautiful day. Showed the garden and arboretum to Pamela Miles.

————

Tuesday, 23rd October.

Went into Bury with Fanny and Pamela Miles. Went to the Quarter Sessions—County business. Pamela was photographed. Visits from Mrs. Wilson and Vice-chancellor Kindersley.† The Abrahams and Lady Rayleigh and her son and daughter arrived.

* The public orator.   † Mrs. Wilson's father.

Captain and Mrs. and Miss Horton came to luncheon and to see the pictures. The Abrahams went away. Sir Edmund and Lady Head arrived: —very agreeable.

Read an article in the new number of the *Edinburgh Review* on the Indian Mutiny. The Reviewer contends (in opposition to Mr. Kaye) that it was strictly a military mutiny—a revolt of the Sepoy army:—not an insurrection of the people generally: that it was not occasioned by any widespread popular discontent, nor provoked by any oppression on the part of our Government: and the principal lesson to be drawn from it is the danger of keeping up a very large native army.

Thursday, 25th October.

Pamela and Philip Miles went away: in the afternoon Minnie and Sarah went off to Riddlesworth. Mr. Clark the public orator of Cambridge arrived, also Mr. Lott. The Abrahams, Lady Cullum and Vice-chancellor Kindersley dined with us.

Here are two good anecdotes,—the one told by Sir Edmund Head, the other by Mr. Clark of Trinity College Cambridge, about Pope Gregory XVI. The first, a Cardinal, being in favour and on familiar terms with His Holiness, went one day into his private apartments, and found him smoking. The Pope offered him a cigar,—"Grazie, Santo Padre, non ho questo vizio, Se fosse vizio l'avresti," replied the Pope. The other was a squib which

1866. was circulated at Rome in the life-time of Gregory.

The Pope is supposed to have died, and applied at the gate of Paradise for admission; St. Peter says to him "You have your own key, you can let yourself in." The Pope takes out a key, tries it at the lock, fumbles for a long time, cannot open the gate, at last the horrid truth strikes him : "What a terrible mistake! I have brought the key of *la cantina* (the wine shop) instead!"

———

Friday, October 26th.

Lady Rayleigh and her party went away in the morning. Sir Edmund and Lady Head after luncheon. Minnie and Sarah returned. Sir Edmund Head said of "Ecce Homo" that it appears to him to be the work of a man who has been sceptical, and who seeks to escape from intellectual difficulties by giving himself up entirely to the enthusiasm of morality.

Edward mentioned a quaint saying of Sir George Cornewall Lewis. When he (Edward) was in Parliament, on one occasion when Fergus O'Connor divided the House on the Repeal of the Union : as those who voted against the motion were going out together, Cornewall Lewis said to him— "What a proof of self-sacrificing patriotism we are "giving in voting thus, when it would be such a "comfort to all of us to be rid of the Irish "members."

———

October 27th.   Saturday

A beautiful day. I shewed the Ferns to Kate,

Henry and Sarah.   Kate went away after luncheon. 1866.
Mr. Lott and Mr. Eddis also went away.   I finished
vol. ix. of Froude, the 3rd volume of the reign of
Elizabeth.   Much of it is very interesting, but some
parts of it are, I must confess, rather tedious, over-
loaded with an excess of detail of the negociations
and of the Queen's waverings and uncertainties.
The history altogether is certainly on a large scale :
this thick volume of very nearly 600 pages contains
the transactions of just three years, from February
1567, to February 1570 : and the following volume,
the 4th, only brings down the history to the 15th
year of the reign.

This 3rd volume belongs almost as much to
Scottish history as to English; Mary Stuart and the
Regent Murray play very nearly as prominent parts
in it as Elizabeth.   Murray, indeed, is Froude's
especial hero : he pronounces him worthy to "take
"his place among the best and greatest men who
"have ever lived."

He gives us rather a new view of the Regent's
murder.   I had been under the impression (derived
indeed mainly from Walter Scott) that it was
principally an act of private revenge.   Froude shows
that it was the result of an extensive conspiracy, in
which the whole of the House of Hamilton was
more or less concerned, together with many other
enemies of the Reformation.

Mary Stuart and Elizabeth appear, on the whole
in pretty much the same colours in this volume as in
the preceding ones.   Froude is very severe upon
Mary, and certainly shows ample reason for it ; but

P

1866. of Elizabeth he gives me, I really think, a more
unfavourable impression than any other historian I
have read. Her continual vacillation and excessive
infirmity of purpose, her niggardliness, the meanness
of her conduct towards all whom she employed,
excite a feeling really approaching to contempt.
She may not have been intentionally treacherous,
but her mean passions and infirmity of purpose,
made her act in a manner, which had all the effects
of treachery, but towards Mary and towards the
Regent, one feels that it was truly owing to an over-
ruling Providence, that her weakness of conduct
was not ruinous to herself and to the Reformation.

Under Providence, both England and the Reformation were saved by Cecil more than by anyone
else.

Elizabeth, however, had some very good servants
besides Cecil : — Sir Francis Knollys, Lord Huns-
don, the Earl of Sussex, Randolph and others,—
though she hardly deserved to have such. She
must have been a most intolerable mistress to serve,
from her incessant changes of purpose, and her
practice of never giving definite instructions to those
she employed, but leaving them to bear the
responsibility of whatever she wished them to do.

There is nothing in this volume of Froude that I
more heartily approve and admire than the beginning of the 16th chapter, where he speaks of the spirit
of Christianity and the opposition between it and the
spirit of dogmatical theology. It is admirable.
The character of Philip II. of Spain is very ably—
though rather too indulgently sketched, as well as that

of the Duke of Alva. The latter appears to me a 1866. remarkably clear, fair and probable portrait.

One thing very new and unexpected—at least to me—and which seems to be quite clearly established by the correspondence which Froude has brought to light, is the remarkable patience and forbearance which Philip II. showed towards Elizabeth in the first ten years of her reign:—his extreme unwillingness to quarrel with her. He endured repeated provocations from her subjects and her ministers, such as even in these days would be thought sufficient to justify war. It is true that this forbearance might not be owing so much to any regard for her, or to any mildness of disposition as to his apprehensions of France, and his fears of driving her into a close union with that government.

Monday, October 29th.

Dear Minnie and Sarah went away. I was very sorry to part with them. Fanny and Edward went with them to Cambridge, spent the day and returned late.

LETTER.

Barton,
November 1st, 1866.

My Dear Katharine,

I am very much obliged to you and to Mrs. Bentham, for information about the Tradescantia *Zebrina*, the name of which I am very glad to know, and it is, as you say, an appropriate name. The

1866. plant is not in the *Botanical Magazine*, at least not in the volumes which I have. You say it is very unlike either a Sedge or a Lily, and so it is perhaps if looked at in an isolated way. But I have seen some tropical grasses which in their leaves and mode of growth, were so like Commelynas (which are next of kin to Tradescantia) that one might at first sight take them for something of that sort without flowers. Then again, as to the flower, one may easily trace a series of connecting links, from Lilium and Fritillaria, through Calochortus, Butomus, Limnocharis and Alisma, to Tradescantia. I should have mentioned, too, that some Tradescantias have leaves quite of the Lily type.

I return the copy of the Address to George Pertz, which is undoubtedly very gratifying.

I am very sorry to hear that your poor Aunt, Mrs. Power, is so ill.

Believe me ever,
Your very affectionate Brother,
CHARLES J. F. BUNBURY.

JOURNAL.

Friday, November 2nd.

Edward went away and we were at last alone.

Tuesday, November 6th.

A beautiful day ; out all the morning with Fanny. We visited several cottages, and settled about planting roses in the cemetery.

Read the article in the last number of the *Edinburgh* on "The Military Growth of Prussia"—clear and instructive.

We went to Mildenhall on a visit to Sir Edmund and Lady Head. Mr. Lott dined there.

Visited Mrs. Buck.

Took a walk with Sir Edmund to the "Park" and Barton Mills—pleasant talk.

We dined in the middle of the day with Mrs. Buck, who was in very good spirits on the eve of her 84th birthday. Visited the boys' school. We returned home.

Went to the gaol committee at Bury.

We discussed the question of the "separate" cells, and Wilson undertook to draw up an answer to the Inspector,

Finished reading "Ecce Homo."

A remarkable work certainly, though I can scarcely understand the enthusiasm which some of

1866. my friends feel for it, and much less the rancour
with which it has been assailed on the other
side.

Sir Edmund Head's remark on it, which I
recorded on October 26th, appears to me very just.
The book is in fact an essay on the *moral* aspects of
Christianity : on the moral teaching of Christ,
(considering his example as an important part of
his teaching) ; it distinctly professes to keep clear of
all doctrinal questions, and it appears to me very
unfair to find fault with it for not containing what it
does not profess to contain—a complete survey of
Christianity. It is evidently the work of a man of
ardent temperament—of a very enthusiastic character ; and sometimes, I think his enthusiasm carries
him too far. This is more particularly the case in
the chapter on the "Law of Resentment," in which
he is so hurried away by his ardour, that he actually
seems to approve of religious persecution in strange
inconsistency with his usual benevolence.

I think too, that, like most writers on the same
subject, he is a little too hard on the ancient Pagan
moralists,—unfair in this respect at least, that he is
apt to compare the theory of Christianity with the
practice of Paganism. I am afraid that if the comparison were fairly made between the practice of
the professors of the two religions (taking periods of
somewhat equal civilization in other respects), the
contrast would not be so striking. I am afraid the
author has rather exaggerated the actual improvement in moral goodness, produced by Christianity.
At any rate to find any striking improvement, one

must compare the unobtrusive course of private life, 1866.
and not the proceedings of governments and statesmen.

LETTER.

Barton,
November 18th, 1866.

My dear Joanna,

I thank you very much for your very kind
letter of the 11th, which has given me great
pleasure. It is a long time since I have written
to you, but I trust you know me well enough to
believe that it has not been from forgetfulness of
you, and that I have not failed to think often of you
and dear Susan. I felt very much for you at the
time of that most sad and astounding calamity in
the Pulszky family, well knowing how much you
were attached to them all, and how great a shock it
would be to you. It was indeed one of the saddest
events within my knowledge : — such a rapid accumulation of sorrows on one unfortunate man, that it
seemed hardly credible.

And so Venice is at last really solidly united and
identified with Italy. It is a grand event. While
the political prospect in other directions looks
gloomy, unsettled and unpromising, this is I hope
a real permanent good that has been gained, and
I trust that the Italians will now direct their
thoughts to peace and industry, and internal improvement ; though I fear they may have some
trouble yet with Sicily and Naples. Fanny's.

1866. journals will have kept you acquainted with our gaieties, and the long succession of company that we entertained from the 21st of August to the beginning of this month. It was very pleasant —several of our parties very agreeable indeed, but in the long run it was rather too much for her, and she was quite knocked up ; almost ill from sheer fatigue.

After Edward left us on the 2nd of November, we were for nearly three weeks quite alone (except a visit of three days to the Heads at Mildenhall), and we enjoyed the rest very much, and it did us— especially her — a great deal of good. Since I began this letter we have had another little fit of gaiety : Sir John Kennaway and his son and daughter staying with us from the 19th to this morning the 22nd ; the Abrahams and Waddingtons being here the first day, the Arthur Herveys the other two ; as also Mr. Lott and Mr. Tyrell. Sir John Kennaway is an elderly gentleman from Devonshire, with whom we made acquaintance at Sir John Boileau's in London, where Fanny took a great fancy to him. Miss Kennaway is handsome and remarkably intelligent and agreeable, and the young man is also very pleasing. The Arthur Herveys we both love dearly, and Mrs. Abraham too.

It is, I think since my letter to you, that I have read Baker's "Great Basin of the Nile," one of the best books of *adventurous* travel, I think, that I have ever read. By saying "adventurous travel," I mean to distinguish it as belonging to a distinct

class of travels from such as those of Humboldt, 1866.
Martius, Ulloa, Darwin, Bates, and many others,
in which the personal narrative is quite subordinate
to the scientific or other information. Baker is a
traveller of the class of Bruce and La Vaillant, and
a remarkably entertaining one; a very clear and
lively writer. The wonderful thing is that his wife
should have accompanied him all through that
tremendous journey. I dare say you will have
heard of her from Mary, who has met them in
London.

I am much grieved that your excellent aunt, Mrs.
Power, is in such bad health.

With much love to Susan,

I am ever your very affectionate Brother,

CHARLES J. F. BUNBURY.

—————

JOURNAL.

Tuesday, 20th November.

Fine, hard frost, very cold. Walked round the
gardens and arboretum with Miss Kennaway and
others of the party. Showed her the Ferns. The
Abrahams and Waddingtons went away after
luncheon. The Arthur Herveys arrived,—also Mr.
Lott:—a very pleasant evening.

—————

Wednesday, 21st November.

This evening Miss Kennaway sang very sweetly.
Lady Cullum dined with us.

—————

We went to luncheon with the Arthur Herveys,
and spent most part of the day there.   Met Lord
Charles and Lady Harriet Hervey, and Lord and
Lady Clancarty.   Lord Charles very agreeable.
Sarah Hervey charming.   A very pleasant day.

Received vol. i. of the new (10th) edition of
Lyell's "Principles," and read chapter 9, which treats
of the "Progressive Development" of organised
beings in geological time.   This is in a great
measure new, and is extremely good.   His treatment
of the subject appears to me particularly judicious,
without committing himself absolutely to the
"Progressive Development" theory, he gives an
excellent sketch of the evidence which we possess
on the subject, and fairly and candidly admits the
preponderance of evidence in favour of the theory.
By the way, it may be advisable to devise some
other term for this doctrine than that of "Progressive
Development," which might readily be confused
with the *Development* theory of Lamarck and
Darwin.   With *that* theory it has no connection.
What is meant by it is, the successive appearance
in the course of geological time, of continually
higher and higher forms of organic life, the appearance of an ascending scale of organic types, as we
advance from earlier to later deposits.   Lyell was
for a long time an opponent of this theory, and it is
the more honourable to his candour that he has now
fully acknowledged the probabilities in its favour.
I quite agree with him too, that the evidence in its
favour from the vegetable world is stronger than the
animal.

Saturday, December 1st.   1866.

Read chapter 10 of Lyell's new edition on the Evidences of Former Differences of Climate, and specially on the Climates of the Post-Pliocene and Tertiary Periods: — excellent. The history of the Siberian elephant is particularly good.

LETTER

Barton,
December 2nd, 1866.

My Dear Katharine,

I am very much obliged to you for your information about the *coloured* copy of "Icones Filicum."* I have written to Willis and Sotheran to secure it, for it is a thing I have been some years looking out for, and is seldom to be met with. As you say, it is inconceivable how it can be sold so cheap. I am very glad you have got a good copy of the Icones. I will bring to town with us the volumes of Hooker's "Journal" which contain Smith's list of Phillipine Ferns ; Sir W. Hooker's of Chinese Ferns, and Schomburgh's Ferns of Guiana.

I shall be very glad to spend an afternoon with you while we are in town, to look over your specimens and compare notes.

Ever your very affectionate Brother,
CHARLES J. F. BUNBURY.

* It was bought at Mr. Lyell's sale at Kinnordy.

JOURNAL.

1866. Went to the Gaol Committee at Bury—Mr. Rodwell, Anderson, Abraham, Huddleston, James Bevan and Phillips—long discussion, satisfactory agreement at last. Selected papers and made various arrangements preparatory to going to London.

Thursday, December 6th.

We went to London. Mary met us at 48, Eaton Place.

Friday, December 7th

We dined with the Henry Lyells. Met Charles and Mary, the Youngs, Charles Lyells and Edward.

Saturday, December 8th.

Fine, but much colder.

We drove out to Wimbledon Common. Saw Emily Napier and her children; also Cecil, who is under orders for Ireland. William Napier was in London.

Sunday, 9th December.

We went to morning Church at St. Michael's; Sir Edmund and Lady Head came to luncheon with us.

Monday, 10th December.

We had luncheon with the Charles Lyells.

We dined with Charles and Mary; met Miss Coutts, Julia Moore, Mr. Lecky, Joseph Hooker, and Edward; very pleasant. Many more in the evening. Bentham, Hardcastles, Romillys, Louis Mallets, Mr. Maurice, Mr. Wallace, &c., &c.

Wednesday, 12th December.

Finished chapter 13 of Lyell's " Principles." New edition, relating to the astronomical causes of geological changes. He here treats of the probable or possible amount of influence which such causes as the precession of the equinoxes, changes in the eccentricity of the earth's orbit, and in the obliquity of the earth's axis, may have had on geological phænomena, particularly on the glacial period. The subject is a very difficult one, ignorant as I am of astronomy. What I do understand, and what he seems to me clearly to have made out. is, that the effect of these astronomical causes must have been entirely subordinate to that of the changes in physical geography and almost insignificant in comparison.

Last night at Charles Lyell's I was introduced to Mr. Wallace, the great naturalist traveller. He said that he did not think the Indian Archipelago richer in variety or beauty of natural productions than tropical America, but it had been much less explored, and therefore afforded more novelties.

His opinion of the Malays was that they make admirable servants if due regard be paid to their feelings and prejudices; they readily become

1866. attached, and are more docile and tractable, more trustworthy and more industrious than the Negro races, but he thought them less intelligent, which surprised me very much.

He says he must prepare for publication the narrative of his travels in the Indian Archipelago (of which he has seen so much more than almost any other European) though he would much prefer working entirely at the Natural History results of them.

At the same party we saw the famous Mr. Lecky, the author of the important work on " Rationalism," a tall, fair, young looking man, of extremely gentle, modest, quiet, indeed retiring manners—not talking much.

Joseph Hooker was there also ; I was very glad to see him looking well, and seemingly in very good spirits.

Thursday, 13th December.

I went to Katharine's, and spent a pleasant hour with her, looking over Ferns, a set of very beautiful and interesting ones, sent to her from Borneo,— some of them apparently identical with those of the Khasia mountains, others different, and some of them very unlike any I had seen before. Some magnificent Lycopodiums of the Selago and Phlegmaria sections.

I also looked over with her, the papers which she has written for her intended work on the geographical distribution of Ferns.

I had luncheon with her, and I then went to the

Athenæum, where I met Sir Edward Ryan; the 1866.
Miss Richardsons, William Napier and Edward
dined with us.

Friday, 14th December.

Went in the morning to Harley Street, and had
a pleasant talk with Charles Lyell.

Katharine read to me part of a letter she had just
received from Lady Smith, Sir James' widow;—at
the age of 94, still writing with the utmost spirit,
intelligence and animation on Lord Sidney Osborne,
the Bishop of Oxford, Baker's "Nile," and other
topics.

Saturday 15th December.

Mrs. Holland,* an old friend of Fanny's, came to
luncheon with us.

We drove out to Wimbledon and dined with the
William Napiers.

Sunday, 16th December.

Charles and Mary dined with us.

Read 16th chapter of Lyell's "Principles:" on
the effects produced by ice in its various forms of
drift-ice, glaciers, icebergs and coast-ice, all described
in a most clear and satisfactory way, especially what
relates to the glaciers. The part that is most new
is the description of the curious glacier lake, the
Märjalen sea, which he visited last year.

I finished Trollope's "History of Florence." The
concluding chapters of it relate that most sad story
—of the famous siege—of the last fruitless, but most

* Daughter of Lord Gifford.

1866. noble struggle of Florence for her liberties. It was indeed a glorious effort when the Florentines alone almost, against the world, surrounded on every side by enemies or treacherous allies, and with enemies and traitors in the midst of them, determined to hold out to the last against the overwhelming forces that threatened them. And it really seems as if they might have succeeded, at least for a time, if their general had not been a vile traitor.

To the last they suffered as they had so often done before, from the mania of trusting to hired commanders.

If Ferruccio instead of Baglioni, had been their General-in-chief, the fall of Florentine liberty might at least have been delayed.

On the whole I like Trollope's book; it appears to me honestly and conscientiously written, with much good feeling and good sense; and his political reflections in general are to my thinking very sound. His style is not to my taste; it is too familiar, often almost bordering on *slang*, and entirely wants dignity; but it is lively and entertaining and easy to read.

Tuesday, December 18th.

We dined with the Youngs, and met the Louis Mallets and Edward.

Wednesday, December 19th.

We went out to Twickenham and spent the afternoon with the Richard Napiers:—as kind and cordial as ever: *he* in bad health, but his mind 1866. quite clear and strong.

A horrible fog. A long visit from Mr. Matthews.* Harry Lyell and his children came to luncheon with us, and afterwards we had a visit from Charles and Mary.

An intense fog which lasted all day. We returned home—all well, thank God. We fell in with Arthur Hervey, Sarah and Kate at Cambridge Station; they travelled with us to Saxham Station.

Walked round the arboretum and gardens, inspected the Ferns, etc. I sent Lyell an extract from the Amœnitates academicæ, of which, perhaps, he may make use in the next volume of his "Principles." It is from the 32nd dissertation of the Amœnitates, entitled *Plantae Hybridae*. What is remarkable in it is, that it shows Linnæus to have been ready to admit, that (in genera which have a great number of species in some one country) many of the species now acknowledged as distinct, may not have been originally created so, but may have originated in whatever manner from previously existing species. He mentions as instances of such genera, the Asters and Oaks of N. America, the Heaths, Mesembryanthemums and Proteas of the

* Who had been his Tutor for many years.

Q

1866. Cape, etc. All naturalists, I suppose, would now agree that he was wrong in attributing the origin of these permanent forms to hybridity ; but this is of minor importance in its bearing on the history of the theories of species. The remarkable point is, that he admits the possibility of species, not originally distinct, having *become permanently so.*

LETTER.

Barton,
December 23rd, '66.

My Dear Lyell,

I send you the extract from Linnæus which we were talking about the other day. It does not go so far, nor express his opinion so clearly as I had thought from looking hastily through the paper; but I think it certainly implies, that he suspected some, or perhaps many, of the clearly allied species of those genera which he mentions to have originated from hybrids, and to have become so far permanent as to require to be treated as distinct species. The extract is from the latter part of the 32nd dissertation of the Amœnitates academicæ (vol. iii. pp. 28-62) entitled *Plantae Hybridae.* In this dissertation, Linnæus describes in detail many examples of what he considered hybrid species (though all botanists of the present day hold him to be clearly wrong in this, as to all the cases); then he proceeds to enumerate several that he considers as suspicious, "*suspectae*," and ends with the passage which I have extracted relative to exotic plants.

The dissertation was delivered, it appears, in the 1866. University of Upsal, on the 23rd of November, 1751. What I think this passage shows, is, that Linnæus was ready to believe that many species of plants, now acknowledged as distinct, might not have been originally created so, but might have been derived (by whatever process) from other species.

With much love to dear Mary,

I am ever your very affectionate Friend,
CHARLES J. F. BUNBURY.

JOURNAL.

Wednesday, 26th December.

A beautiful morning. We went to see the Christmas-tree at the girls' school. Had a visit from Lord Arthur Hervey and Sarah.

Sunday, 30th December.

Finished reading Miss Eden's " Up the Country," a re-publication of letters written to her sister in England, while she was accompanying her brother Lord Auckland on his official tour through the north-western provinces of India. It is very lively and entertaining; written in a remarkably pleasant style, easy and yet sparkling; but it conveys very little information. She seems to have been thoroughly out of her element in India; to have felt all its disagreeables most keenly, and to have cared nothing for what is really interesting in it—

1866. to have been always longing for London and for London society.

We went to morning Church and received the Communion.

_____

Monday, 31st December, 1866.

A very beautiful day—very bright and still; took a long walk with Fanny, she riding.   Visit from Lady Cullum.

Finished chapter 20 of Lyell's " Principles," on the effects of tides and currents.   This chapter has always been an especial favourite with me, ever since I first read the first edition ; the description of the inroads of the sea on the English coasts followed in detail all along the coast from the Shetlands to the Land's End, is remarkably good.   The principle novelty in this edition consists in the particulars relating to St. Michael's Mount.

The servants' dance.

_____

So ends 1866.   Thanks be to God for all the blessings enjoyed in it.

_____

DESULTORY NOTES.

I suppose it is true that during the years of childhood one learns more, one receives a greater number of new impressions and imbibes a greater number of new ideas ; the development of one's mind proceeds at a greater rate than in any other period of life. But this development is in great part unconscious ; one does not observe and cannot measure the wealth of new ideas added to one's mind in childhood.

Speaking of what I *know*, and am conscious of, I can say that the period of life in which I gained the most, intellectually—in which my mind was most enriched, grew to its full stature, was between the ages of 18 to 30.

In these 12 years were comprised :—

Firstly, my first acquaintance with foreign countries, in the tour which I made with my Father and Mother and Brothers, in France and Italy, and after my Mother's death, with my Father and Brothers, in other parts of Italy and in Switzerland. It is difficult to exaggerate the quantity of new food that is supplied to one's intellect, when one first goes out of one's own country, under the direction of so powerful and so well-stored a mind as my Father's. He was as well qualified, as he was well disposed, to guide my curiosity, and direct my observation to the objects most deserving, and to guard me against that dissipation of mind of which

1866. there is so much risk when one is thrown without a guide amidst a multitude of new objects.

Secondly, my experience at Cambridge, when I learnt something from the regular course of study, and more from some of my companions, — and perhaps most from "the Union."

Thirdly, my first serious thoughts on politics, much assisted by the careful reading of Hallam's "Constitutional History," which I should consider one of the most important books in our language.

Fourthly, the first reading of Lyell's "Principles of Geology" and Lindley's "Natural System of Botany," two books which made a great impression on me, and gave me the first idea of the philosophy of the natural sciences. Both these I read while at Cambridge.

Fifthly, my visit to Brazil, by which indeed I did not profit so much as I ought to have done, but which, nevertheless stored my mind with a multitude of beautiful and glorious images.

Sixthly, my frequent and long conversations with Sir William Napier, whom I used to visit annually, when he lived at Freshford. I could not fail to learn much from this intimacy with a great genius, and a noble though singular character, whose talk was copious and unreserved, and who, however erroneous one might think some of his views, never failed to expound and to maintain them with wonderful vigour and eloquence. He loved paradox and discussion, and his powerful though not always accurate memory supplied him with abundant materials for defending his opinions. He made a great impression on me.

Seventhly, my canvass and contest for the borough of Bury, which added something to my knowledge of mankind.

Eighthly, my stay at the Cape with Sir George Napier, one of the most interesting and most profitable years of my life. In this outline I have noticed only those things that contributed to my intellectual progress, not those of which the effect was chiefly moral.

1866.

C. B.

(Written in September.)

———

"Plutarch's Lives," is a book which I think particularly useful for boys and young men, and which I would always put into the hands of a boy, feeling sure that if there were any elements of nobleness in his character, it would call them forth; and that it would do him nothing but good. It was my delight when I was about 11 to 13 years of age (1820-22). I read the "Lives" over and over again. I knew them almost by heart. They did not indeed excite me to the same violent degree as they did Alfieri (see his life), but few books made a deeper or more lasting impressions on me; and there are few to which I feel myself more indebted. I am sure that many men must have owed to that book, their first lessons of magnanimity; their first ideas of great and heroic characters. Therefore I think it should form an essential part of the studies of every English gentleman; and that it is more important in this view than many books

1866. which are preferred on account of their superior merits of style.

I read Plutarch in Langhorne's translation, which I believe is not now highly esteemed ; but from the association of ideas with those early studies, I cannot now read him in any other.

1867.

1867.   During January we were quite alone, the cold was very severe.   Whenever it was fine enough we took walks together.

The books he read during this time were as follows:—

Sir Charles Lyells "Principles of Geology," (new edition).

Bacon's "Atlantis."

Ovid's "Fasti."

Kaye's "Sepoy War."

"I. Miei Ricordi," by Massimo d'Azeglio.

F. J. B.

JOURNAL.

January 1st, Tuesday.

A considerable fall of snow in the night, ground covered with snow which remained all day with hard frost—a clear bright day.   Examined accounts with Fanny.   Worked at my biographical notes,

and went on with the catalogue of my herbarium. 1867.
Read chapter 21 of Lyell's "Principles."

Very hard frost and more snow. Read Mr.
Churchill Babington's introductory lecture (at
Cambridge) on Archaeology: he sent it to me the
other day. It is interesting.

The labourers' supper party.

Read Darwin's very remarkable paper in the
Journal of the Linnean Society, on the three
different sexual forms of the Orchid Catasetum
tridentatum, which have been taken for three
distinct *genera*. He shows that the so-called
Catasetum tridentatum is the male, Monachanthus
viridis the female, and Myanthus barbatus the
hermaphrodite form of the same species. I heard
this paper read before the Linnean Society at the
time. I have made a note of it, in the blank pages
at the end of Humboldt's "Aspects of Nature."

Read chapter 2 of D'Azeglio's "Ricordi." The
account he gives of his father and mother is quite
charming.

Went on with my biographical notes. Studied
Göppert's Fossilen Coniferen. Read Ovid's "Fasti."

1867. We saw in the *Times* the account of the sad disaster on the ice in Regent's Park.

———— —

January 17th.

More snow. I received a letter from Katharine, which relieved our anxiety about her boys.

LETTER.

Barton,
January 17th, 1867.

My dear Katharine,

I was very thankful to receive your letter this morning, and to learn that your dear boys were safe. It is indeed a mercy. The account of that terrible disaster in the Regent's Park made me feel very anxious about them, though it is always my natural tendency in such cases rather to hope the best than to fear the worst ; and if I was anxious, you may imagine what Fanny was. God be thanked for their safety ; I do not know when I have felt more thankful. But there must be a sad number of families in grief and distress.

I have not time to write more just now, only to thank you very much for your letter.

The snow is so heavy. It is long since I have seen such a regular old-fashioned snowy winter.

Ever your very affectionate Brother,
CHARLES J. F. BUNBURY.

Charles and Mary arrived safe and well.

———

January 22nd.

Hard frost still.　Fanny and Mary went over to Mildenhall, returning between four and five.

———

January 23rd

Talk with Charles and Mary.

———

January, Thursday 24th.

A beautiful day ; the snow nearly all gone.　I had a walk with Charles Lyell.　Resumed the catalogue of my herbarium.

The Abrahams, Mr. Lott, George Napier, Cecil, and Herbert arrived, and Lady Cullum came to dinner.

———

January 25th.

I had a good walk with Charles Lyell, Mr. Abraham and Mr. Lott.　The Arthur Herveys dined with us.　I had a talk with Sarah.

———

January 26th.

The Abrahams and Mr. Lott went away.　I took a long walk with Fanny (who rode on Topsy), and Charles Lyell.　The Louis Mallets and Georgina Pellew arrived.

———

I had a walk with Mary. We had a large dinner
party:—Lady Cullum, the Wilsons, Gilstraps, Mr.
Borton and John Phillips.

————

January 29th.

George went away. I had a walk with Charles
Lyell and the Mallets. Charles Lyell said, that
nothing contributed more to shake his belief in
the old doctrine (which he formerly held) of the
independent creation of species, than the facts of
which so many have lately been recorded, relating
to the rapid naturalization of certain plants in
countries newly colonized by Europeans. He
remarked that these introduced plants, many of
which have spread to an enormous extent and with
surprising rapidity in the Australian colonies, New
Zealand and parts of S. America, belong in many
cases to families entirely wanting in the indigenous
floras of the countries in which they have thus
settled themselves, and hardly ever to families
prevailing in or characteristic of those indigenous
floras. When one sees, he said, a particular genus
or order of plants abounding very much in a
particular country and exhibiting there a great
variety of specific forms, one is naturally inclined to
suppose (on the "independent creation" hypothesis)
that there are particular local conditions in that
country, which render it peculiarly suitable and
favourable to that family of plants. But when we
see an introduced stranger which has no affinity to
that prevalent family, intruding itself in its place,

overpowering and superseding it, this explanation 1867. becomes less satisfactory, and one is led to search rather for some law of descent with variation, to explain the multiplicity of nearly allied forms in a particular region.

Lyell thinks very highly of the Duke of Argyll's book "The Reign of Law," and says that he has combated Darwin on some points with great force and justice, and has well-exposed the weak points of his system. Where, however the Duke attempts to deal with the great abstract questions of moral Liberty and Necessity, Fate and Free Will, he is not more intelligible or satisfactory than the generality of writers on those questions. The Natural History part of the subject he understands, and upon that he is very strong.

Lyell agreed with me that Darwin is too apt to exaggerate the importance of his hypothesis of Natural Selection, to *deify* Natural Selection (this was Lyell's expression) to speak as if Natural Selection were a great primary law of nature, which would explain the real origin of all the diversity of organic forms : instead of being at the utmost, the process by which varieties are segregated into species. He is also, Lyell said, too scrupulous in avoiding any but the slightest admission of, or allusion to a first cause: even avoiding with excessive care any reference to a Designer, while (as in his book on Orchids) he continually points out proofs of Design. This dread (as it were) of any reference to primary causes seems to be owing (Lyell said) to a reaction against the too great readiness of some of the

1867. naturalists to refer on every occasion to such causes,
thereby saving themselves the trouble of investigating
secondary causes.

Agassiz (Lyell says) runs into great extravagancies
in this way, maintaining not only that all species
were created separately, such as they now are, but
that whenever a species exists in two distinct parts
of the world, it was separately created in both; and
moreover that there was an entirely new creation
at the beginning of every geological period.

He holds, in short (like Uncle Toby concerning
noses), that there is no reason why one species
resembles or differs from another, except that the
Almighty made them so. Undeniably true, no
doubt, and saves all the trouble of philosophizing.

Lyell told me that a stranger once sent him a
drawing of what he supposed to be a fluted column,
very skilfully wrought and elaborately ornamented,
and which, he added "proves the antiquity of man
to be much greater than even you have hitherto
represented it." It was a fine Sigillaria, of the
coal formation. And the man had no ironical
meaning, but quite seriously believed it to be a work
of art. Lyell remarked that when we read Linnæus's
writings in such clear intelligible (though peculiar
Latin), we feel very sorry that the Swedes, his
countrymen, should have lately taken to writing
their scientific works in their own language, which
makes them sealed books to most parts of the
world.

Lyell is much pleased at Lord Stanley's pro-
posing to refer the "Alabama claims" to arbitration.

He said that Lord Russell's reply to the American 1867.
Government on that subject was too much in the
spirit of a duellist in the old days.

He is very angry with President Johnson,—among
other things,—for recalling Mr. Motley from the
Embassy at Vienna.

Charles and Mary went away, very sorry to part
with them. We dined with the Arthur Herveys—
met Lord and Lady Bristol and Lady Mary. Mrs.
and Miss Harvey from Bedfordshire, and Mr.
Bevan.

I took a walk with Fanny, Mr. Lott and the
Mallets. We went to the amateur concert at the
Athenæum at Bury, — very successful — delightful
singing by Mrs. Abraham, Mrs. and Miss Harvey,
and Lady Augustus Hervey.

LETTER.

Barton,
February, 1st, 1867.

My Dear Mary,

We were both very happy to hear of your
safe arrival. I am writing for Fanny, because she
has been quite knocked up by the concert at the
Athenæum last night. The concert was a grand
success, and very well managed in every way; I
really enjoyed it very much. Mrs. and Miss

1867. Harvey (friends of the Arthur Herveys, from Ickwellbury, in Bedfordshire) and Mrs. Abraham sang delightfully, and Lady Augustus Hervey wonderfully. The room was full, and the audience very enthusiastic. I ought to have mentioned Maitland Wilson, who was very good too, as well as some others whom I am not acquainted with.

In my youth, how I should have been astonished at the notion that I could enjoy a concert.

The Mallets went away early this morning, and Mr. Lott after breakfast, and we are now alone with our two Nephews.

Dear Mary, I very much enjoyed the time that you were here, and the conversations with you and Charles Lyell. It is always a great pleasure to have you with us, and I trust the change did you good. I hope you will both take great care of yourselves.

I see in the papers the death of Lord Camperdown, whom I knew well at Trinity. He was four years younger than me, but was of the same standing at Cambridge.

Pray give much love from me to dear Katharine as well as to your husband, and believe me,

Your very affectionate Brother,

CHARLES J. F. BUNBURY,

JOURNAL.

Monday, February 4th.

My 58th birthday. Thanks be to God.

Received a kind letter from Cissy. Wrote to her and to Henry.

Tuesday, February 5th. 1867.

Began to unpack and arrange my collection of fossil plants, brought from Mildenhall.

====

LETTERS.

Barton,
February 5th, '67

My Dear Katharine,

I thank you very heartily for your most kind letter and all your good wishes, as well as for the very pretty slippers which I received yesterday morning, and which I prize very much. I am very sorry to hear that you are suffering from neuralgia. I have been quite well since the severe weather departed,—as well as I ever am in winter.

We were shocked by hearing yesterday morning that Mrs. Bucke had fallen down stairs and broken her arm; it appears to have been well and speedily set, and the accounts this morning both from Miss Bucke and Mr. Image were as comfortable as could be under such circumstances; I hope she will have vital power enough to get well over it, but such a shock at 84 is a very serious matter.

I wrote Mary some account of the concert at Bury, which was extremely successful, and I dare say she will have told you about it. We have no more gaieties definitely in prospect before we leave home, but I hope Kingsley will come to us before long. I have at last begun to unpack and re-arrange here my collection of fossil plants, brought over from Mildenhall, which has been lying for

1867. years in packing cases, but I suspect that I shall not have room for them all.

I am reading Kaye's "Sepoy War" (the history of the Indian Mutiny); it is very interesting, and makes me feel that the Indian people (though not perhaps the Sepoys themselves) had many good grounds of complaint. It seems strange that the English, themselves so aristocratic, should have systematically pursued in India a levelling policy—a system of debasing and destroying the native aristocracy. I was much interested by your account of your party and your occupations, and I quite agree with you about Humboldt's Narrative. Are you reading it in French? The English translation does no adequate justice to the delightful style. You will be glad to hear that none of the Coniferae here, and very few of the other plants appear to have been at all hurt by the severe weather. The snow was hardly gone before violets were in flower in the beds and daisies on the lawn. Now the snowdrops are in full blossom, and the crocuses and hepaticas beginning.

Believe me ever,

Your very affectionate Brother,

CHARLES J. F. BUNBURY.

Barton,
February 14th, 67.

My Dear Katharine,

I have got my collection of fossil plants in some degree arranged, and all the more choice specimens very well accommodated in the drawers

of a very good cabinet, but I have much more 1867.
difficulty than I had at Mildenhall in finding room
for large and cumbrous specimens, such as
Sigillarias. The *long gallery* was invaluable for such
purposes.

Fanny had a *Valentine* this morning,—a capital
portrait of Lady Cullum's Garry. I am afraid
there is no improvement yet apparent in Herbert's
hearing. He is a nice boy, very amiable and
intelligent, and Fanny takes a great deal of pains in
reading with him.

I have nearly finished the first volume (a very
thick one) of Kaye's "Sepoy War," and it has
suggested a painful doubt, whether we have any
business to be in possession of India :—whether we
can ever understand the people sufficiently, or
harmonize with them sufficiently to make our
Government really beneficial to them. Mr. Kaye
does not himself draw this inference, but he speaks
of "differences of race, differences of language,
"differences of religion, differences of customs,
"all indeed that could make a great antagonism
"of sympathies and of interests, severing the
"rulers and the ruled as with a veil of ignorance
"and obscurity." This is distressing. However,
as we have got it, I suppose we must keep it,
for I do not imagine they would be at all benefited
by exchanging our rule for any other.

With much love to your children and husband.

Ever your very affectionate Brother,

CHARLES J. F. BUNBURY.

R 2

JOURNAL.

1867.   A beautiful spring day. I was much out-of-doors.
Mildenhall estate accounts of last quarter.   Went
on with my biographical notes.

News of the attempted rising at Killarney.

I finished Kaye's "Sepoy War," that. is the
first volume, all that is yet published. It is an
interesting and important work. It might, indeed,
have been a little condensed, being in parts rather
diffuse and lengthy. It brings us only to the very
beginning of the mutiny, not including any particulars of the outbreak at Meerut or the seizure of
Delhi, but only the arrival of the news at Calcutta ;
nearly the whole of the 618 pages being occupied
by a review of the antecedents and supposed
causes of the revolt.

Mr. Kaye's theory of the mutiny—or rebellion—
as I gather, seems to be this :—that it originated in
the deep and wide-spread discontent, sense of
wrong, and fear of worse, which was excited in the
native princes by our system of annexations, and
in the native aristocracy by our system of levelling
—both these affecting Hindoos and Mahometans
equally ; while in the Brahmin priesthood, there
was in addition, the fear of the loss of their influence through the introduction of European
knowledge and inventions. That these several discontented classes, unable to injure us by their own
strength, applied themselves in various underhand

ways to sap the fidelity of our Sepoys. How the 1867. wild and furious panic about the danger to their religion was created in the native army, Mr. Kaye seems to acknowledge is a mystery. It seems certain that the incident of the greased cartridges was merely the spark which fired the train. There may have been imprudence in some of the authorities, but on the whole, it may be safely said, that neither the British nation nor the Government had the slightest thought of committing that wrong, the apprehension of which drove the native army into frenzy.

All this suggests a painful doubt, whether we have any business to be in possession of India ;— whether we can ever understand the natives sufficiently or sympathise with them sufficiently to make our dominion really beneficial to them ? It is not merely the difference of race and of language ; not merely the usual repulsion and soreness of feeling between conquerors and conquered ; — but the differences of religion and even of moral feeling, of long-established customs and habits of thought, are so extreme, that it is difficult to find any common ground to start from. The whole state of feeling which gave importance to the "greased cartridges," appears so like insanity, that one almost despairs of ever understanding the people who could be influenced by it. Indeed all the notions of "ceremonial impurity" of permanent religious and moral pollution, produced by the contact of something external and material—appears to me to place a formidable barrier between the people

1867. holding such belief and the rest of the world. It is
not as if we had to deal with a barbarous people;
here we have an elaborate civilization of extreme
antiquity and a social system more artificial and
complicated as well as more ancient than our own.

I understand better from Kaye's history than
I ever did before, why Lord Dalhousie was so much
admired by Anglo-Indians in general. He was
(so Kaye relates him) a man of clear views and
intensely strong will, imperious and despotic by
nature, but a benevolent despot; he saw very
clearly what he conceived to be for the good of the
natives, and he was determined to carry it into
effect, whether they liked it or no. He had made
up his mind that it was for the ultimate good of the
people that all India should, as fast as possible,
be brought under the direct dominion of the British,
that all the protected or semi-independent States
should be absorbed; and that everything should be
done to assimilate the population of India to that
of Britain.

Having convinced himself that this course would
be advantageous, he set himself with the most
determined energy to carry it out, indifferent to the
interests and feelings that might be hurt in the
process.

Monday, 18th February.

I have been going on pretty steadily with the
Memoir of my Father's Life; and also by degrees
with the catalogue of my herbarium.

I began to make a new catalogue, much fuller than the old one, of my collection of fossil plants.

We dined at Hardwick—met Sir John Walsham ; then went with Lady Cullum to the Bury Athenæum, and heard Arthur Hervey's excellent lecture on Charlemagne.

Saturday, 23rd February.

A beautiful day. Strolled out with Fanny, and we gave various directions about paths, &c.

Tuesday, 26th February.

Herbert went away. Scott gave a very good account of the rent audit at Mildenhall. I took a walk with Fanny. Went on with my regular pursuits.

Thursday, 28th February.

My Barton rent audit very satisfactory.

Weather fine but cold.

Sent off the last instalment of my donation of £500 to the Albert College.

Went on with my studies.

Saturday, 2nd March.

Lady Cullum dined with us. Kingsley arrived late:—a pleasant talk with him.

Monday, March 4th.

Kingsley went away early.

Tuesday, 5th March.

Wrote to the Secretary of the National Portrait
Exhibition about my pictures.

LETTER.

Barton,
March 6th, 67.

My Dear Katharine,
          Kingsley spent last Sunday with us. He
was very agreeable as he always is, but seemed
overworked and fagged, though in better health,
essentially, than he was last year.

This has been a strange day ; furious storms of
hail, sleet and *snow*, with bright sunny gleams
between. The morning was very bad, and the
eclipse was to be seen only in passing glimpses
through the midst of the drifting clouds ; but even
so it looked very remarkable. We are all in the
"catalogue line :"—you working at your geographical
list of Ferns : Fanny at the library catalogue : and
I have as many as three lists in hand, one (for you)
of the Brazilian localities of Ferns known to me ;
one of my herbarium, and one of my fossil plants.
I am certainly inclined to think that Sir W. Hooker
carried the reduction of species very far, and I am
often at a loss to know on what principle he reduced
some species to varieties, and left others to stand as
distinct, which do not appear more different ;
moreover, I think he was not sufficiently careful in
noting the characters of what he considered as
varieties. It may not be of much consequence

whether a certain well-marked form is or is not 1867. admitted to the honours of a species : but it is important that its distinctive marks, whether specific or not, should be clearly pointed out. I hardly know any book in which varieties are treated so satisfactorily as in the last published volume of De Candolle's "Prodromus," containing the Oaks.

The time is coming near when we had planned to go to London, and as usual I feel excessively unwilling to leave home. The daily routine of our life here is so pleasant, that it makes me very slow to move. But I shall be very glad to see you and a few other people, as well as to hear something of what is going on in the scientific world : and perhaps it is not good for one to be uniformly happy and comfortable.

The poor man, Henry Last, in whom Fanny has been taking so much interest, is going on well, and there really seems good hope of his recovery.

Much love to all your home party.

Believe me ever,

Your very affectionate Brother,

CHARLES J. F. BUNBURY.

JOURNAL.

Thursday, 7th March.

News of Fenian outbreak in Ireland, and of the murder of Dr. Livingstone in Africa.

Tuesday, 12th March.

Engaged most of the day at Bury : first at the

1867. Quarter Sessions, where there was no specially important business, then at the Hospital, where we had a long debate as to appointing another physician.

Wednesday, 13th March.

We arranged books. Began the arrangement for sending pictures to the loan Exhibition at South Kensington.

Monday, 18th March.

The five portraits were sent off for the South Kensington Exhibition, and I wrote to the Secretary about them. Made various arrangements preparatory to leaving home.

Monday, 25th March.

Mild weather, beautiful morning, afterwards rain, We left our dear home and came up to London, arrived safe, thank God. We saw Mary. We heard of poor Mrs. Richard Napier's dangerous illness.

Tuesday, 26th March.

A fine day. We went to luncheon with Charles and Mary. Met Katharine and Arthur. Some talk about the strike on the Brighton railway; agreed that the pretensions of the men are quite unjustifiable; that their demands as to wages and time may have some reason in them, but that the pretension of the men on strike to regulate exclusively the employment and promotion of all on the line, is such as no employer can possibly yield to.* I find that Charles Lyell, though a democrat, is not at all inclined to be a Socialist, or to allow that the working men have alone the right to fix the terms of labour.

I went to the National Gallery and saw the new Rembrandt, "Christ blessing little Children,"—an impressive and interesting picture.

Douglas Galton told me that he is at present employed (besides all his other employments) on a committee which is conducting a series of interesting experiments on the relative strength of iron and steel,—to determine how far, and in what cases, the superior strength (or tenacity of steel) may make it worth while to use *it* instead of iron ; for bridges, for instance.

Met at the Athenæum besides Douglas Galton, Matthew Arnold and Louis Mallet.

Wednesday, 27th March.

Met at the Athenæum Sir Edmund Head, John Moore and Boxall. The Miss Boileaus and Edward called on us. Mr. Boxall said to me that the new Rembrandt in the National Gallery will gain upon me more and more the oftener I see it. Its especial characteristic he says is moral pathos, and this seems to me very true from the one view I have had of it. The face of Christ, though entirely wanting in apparent beauty or dignity, has a very pathetic character. Rembrandt (he says) was an

---

* This strike seems (March 29th) to be already at an end.

1867. *extreme* Protestant, perhaps an Anabaptist or inclining to it, and he may probably have wished to make his representation of Christ as remote as possible from the type received in Roman Catholic Art.

Thursday, March 28th.

We went to the Drawing Room at Buckingham Palace with Charles and Mary. Fanny looked extremely well in her Court dress. Visit from William and Emily Napier: very glad to see them.

30th March, Saturday.

Dear Cissy and Emmy arrived. Emmy grown very tall. The Louis Mallets, Katharine and Edward dined with us. George* came in the evening, he brought the sad but not unexpected news of Mrs. Richard Napier's death.

Wednesday, 3rd April.

Called on the Benthams, and saw them. Bentham tells me that more mischief has been done at Kew, especially among the Conifers, by the frost of this last January, than by the winter of 1860-61. I am very glad to hear from Bentham, that Government have bought Sir William Hooker's herbarium, &c. for the nation, giving £7,000 for it, that is, £5,000 for the herbarium, and £2,000 for the library and other things.

General George Napier.—(F. J. B.)

Lyell told me of a saying of Lady Cranborne,* 1867. immediately after the resignation of her husband and two other Ministers, some one asked her jokingly, whether she had taken a share in the laborious calculations in which Lord Cranborne had been engaged;—No, she said,—the only calculation I have made is a very strange one, and yet very true:—take *three* from *fifteen* and *nothing* remains.

———

Thursday, 4th April.

The Linnean Society Meeting—an important paper on the Geographical Distribution of Ferns was read, by Mr. Baker, a young man who, as I understand, is the assistant curator of the herbarium at Kew, and has had the charge of editing Sir William Hooker's Posthumous Synopsis Filicum. He has therefore had the greatest possible materials and opportunities for such a work, and he appears to have done it, thoroughly. The paper is evidently a very elaborate one, and will no doubt be a very interesting one to study thoroughly, at one's leisure when it is printed; but it is such a mass of statistical detail, such an accumulation of figures, that I could not learn very much from hearing it rapidly read, and still less could I criticize it. His conclusions seemed to be in general, much the same as I had come to, as far as my means enabled me; but I think his limitation of districts questionable. He remarked, what has often struck me, that the Ferns of South Africa have little peculiarity in them, few species peculiar

* Afterwards Lady Salisbury.—(F. J. B.)

1867. to that region, and no group, while the flowering
plants of the same region have a remarkable amount
of peculiarity.—There was not much discussion.
Joseph Hooker unfortunately was absent, having
been summoned as a juror to the Paris Exhibition:
I was much disappointed.   I had a talk with
Bentham.

———

Saturday, 6th April.

Went to the Geological Society.   Called with
Cissy on the Arrans, and then on Sir John Bell also,
and George and Edward dined with us.

———

Monday, 8th April.

Read Horace's "Epistles," 11 to 14, also Milton's
Tractate on "Education," and a part of his
"Areopagitica."

Cissy went to Mrs. Napier's funeral.   Frederick
and Augusta Freeman dined with us.

———

Tuesday, 9th April.

Miss Frederica Johnston and George Napier
came to luncheon with us.   Sir John Hanmer and
Miss Bethel called.   Read Horace's " Epistle," 15
and 16, and some more of Milton's "Areopagitica."

We dined at the Charles Lyells.   Met Katharine,
the John Moores and Sir Edward Ryan.   A party
in the evening.

———

Thursday, 11th April.

The 23rd anniversary of our engagement.   Thanks
be to God.

Read (in the *Morning Post*), the charge of the 1867. Chief Justice to the Grand Jury in the prosecution of Colonel Nelson and Lieutenant Brand, for the proceedings in Jamaica—an elaborate, weighty, and well-considered discourse; of great importance as it appears to me.

In the afternoon we learnt that the grand jury have thrown out both the bills against Colonel Nelson and Mr. Brand; in opposition to what appeared the evident drift of the Chief Justice's charge.

I cannot help thinking that they have done wrong —that it would have been better that the great and difficult questions involved should have been thoroughly argued and examined in open court. But the prosecution has not been entirely useless, since it has produced the very masterly disquisition of the Chief Justice.

I finished reading Milton's noble "Areopagitica," a speech for the liberty of "unlicensed printing." The style is on the whole rugged, the construction often much involved, often more Latin than English and sometimes not free from obscurity; but the grandeur of the sentiments, the noble love of knowledge and liberty that pervades the whole, the force and majesty of the language, make this a wonderful performance.

The picture of the "wealthy man addicted to his pleasure and his profits," who finds religion too hard to understand for himself, and therefore keeps a godly man in his household to be a living religion to him, is one of the most powerful pieces of satire I have ever read.

1867.   We went to the National Gallery and had a good
view of the pictures.   Edward dined with us.

Friday. 12th April.

I received the present of a beautiful drawing of
Mildenhall from Lady Head.   George Napier dined
with us.

Saturday, 13th April.

Lady Bell came to luncheon with us.   The
Ministry, it seems, have been victorious by a
majority of 21 ; in the great and critical division
in committee, on their (so-called) Reform Bill.   I
hardly know how I should have voted if I had been
in Parliament.

I read chapter 2 of Burton's "History of
Scotland," treating of the remains of the Roman
occupation of the country :—very clear and in-
structive.   I went to General Sabine's reception (as
President of the Royal Society) at Burlington
House.

Met Bentham : — he talked of the *Genera
Plantarum*, in which he and Joseph Hooker are so
deeply engaged, and in particular, of the Melas-
tomaceæ, which they have lately finished, and which
employed them a long time, the genera being very
difficult to define.   That is to say, it is very
difficult — even more than usually difficult — to
throw them into generic groups which shall be
natural, and at the same time distinguished by
definite and positive characters.   He said that De
Candolle in his revision of the family, has established

genera without characters : while in Naudin's, there 1867.
are *characters without genera* (meaning definite
characters which artificially separate closely allied
plants).

Read two fine sermons, the second and third of
Kingsley's new volume " The Water of Life and
other Sermons," both vigorous and impressive.

Charles and Mary dined with us—very pleasant.

I received the lease of the mill at Mildenhall and
read it through.

Mr. Weight came and witnessed my signature to
the lease of the mill at Mildenhall.

Visits from Sir Edmund and Lady Head separately.

George and William came to luncheon with us.

Read another sermon of Kingsley's, the 4th of his
new volume—" The Wages of Sin."

I went to the Geological Society ; rather an
interesting meeting. Etheridge read, or rather
spoke, a very elaborate paper on the " Devonian
System," in opposition to Jukes's.* Having carefully
and minutely examined that particular tract of
country in North Devon, on which Jukes had

----

* See Jukes's views in my Journal of March 7th and 29th of this year.

1867. theorized, he maintained the correctness of the classification established by Sedgwick and Murchison, contended that there was no such fault and no such repetition of the strata as Jukes contends for, that the succession of the strata is clearly made out ; and above all that the organic remains (of which he exhibited copious and elaborate tabular lists) distinctly prove the soundness of Sedgwick's and Murchison's doctrine. Much of the argument depends on the accuracy of the sections which can scarcely be judged without accurate local knowledge, but certainly the weakness of Jukes's case, and the strength of the Devonian theory lies in the organic remains, which Jukes appears to have neglected.

Murchison, Phillips, Austen and Ramsey spoke, and with much animation, but there lacked opposition, for neither Jukes himself was present, nor anyone to take up his cause ; it was a debate all on one side ; a sham fight.

Phillips's was very good, vindicating the right of free discussion even of the most established theories, pointing out the real difficulties of the question, and suggesting that the differences of opinion as to what *are* species might have some influence on the comparative lists of Carboniferous and Devonian.

He remarked (what appears to me to have an important bearing on the question) that in Herefordshire and Shropshire the lowermost beds of the true old red sandstone alternate, and are in a manner intermixed with the top beds of the Silurian (the Ludlow rocks) while on the other hand, in parts of the north of England, the *uppermost* beds of the same old red sandstone seem to pass in like manner 1867. into the lowest of the Carboniferous.

--- ---

Thursday, 18th April.

A mild, relaxing day.   A visit from Miss Kinloch —a charming girl.

The third and fourth chapters of Burton's "History of Scotland," relating to "the unrecorded ages," as he calls them—are very curious and instructive, comprising in a moderate compass and in a clear easy style, a great deal of information which was quite new to me about the Scottish antiquities of which the age and history are uncertain.

Mr. Burton is a very sceptical antiquarian, very far from enthusiastic, and much disposed to be satirical upon the eager theorists who have hitherto chiefly dealt with these subjects.

--- ---

Good Friday, April 19th.

I read a most striking and admirable sermon of Kingsley's:—"The Shaking of the Heavens and the Earth."   He treats of the general shaking and breaking up of old beliefs, old opinions and old systems, which is going on everywhere in the present day; the general agitation and disturbance of old notions, not only in the natural sciences, but in politics and religion; and he argues in a noble tone of thought and feeling, that we ought not to look with fear or horror on this general tendency but to believe that it is part of God's plan for the government and education of the world, and that as such it must lead to good.

1867. Katharine came to us and we took a drive with her in the open carriage round Hyde Park, then to her house where we saw Harry Lyell—then we called on Madame Ernest de Bunsen.

Saturday, April 20th.

We called on Lady Rodney, and afterwards to Hobart Place and saw dear Minnie and Sarah just arrived.

Read "Horace;" epistle, 17 and 18, and chapter 5 of Burton on the "Early Races of Scotland." His quietly satirical analysis of the "Pictish Controversy" is very amusing.

April 22nd.

Easter Monday.

Clement went to the Volunteer review at Dover and returned in the evening.

Minnie and George Napier dined with us.

April 23rd.

We dined with the Evans Lombes, and met Baron and Mrs. Marochetti (very pleasant) Sir George and Lady Baker, Colonel Keppel, Sir James Elphinstone and others. Mrs. Lombe very agreeable.

April 24th.

We heard of the death of my cousin John Napier and of the birth of William's eighth daughter.

Minnie and Sarah and George and William Napier dined with us.

I sent a contribution to the Bishop of London's 1867.
fund.

Read chapter 8 of "Early Christianity in
Scotland," and the preliminary or introductory
part of the work. The narrative part begins with
the 9th chapter, but as I am not curious about the
history of the barbarous ages, I mean to proceed at
once to the struggle for independence in the 13th
and 14th centuries.

April 25th.

Dear Cissy and Emmy went away.

Friday, 26th April.

We dined with the Boileaus — met General
Boileau, Mr. George Elliot, Colonel and Mrs.
Stuart. The Miss Boileaus very pleasant.

Saturday, 27th April.

We called on Miss Richardson,* and took her a
drive with us in the open carriage. She was
exceedingly agreeable.

We went to Veitch's garden, and saw beautiful
ferns and flowers.

Clement went away and we were quite alone in
the evening.

Monday, 29th April.

Our dinner party ; Sir Edmund and Lady Head,
Mr. and Mrs. Matthew Arnold, Mary Lyell, Miss
Richardson, Sir George Young and Mr. Boxall.
Minnie and Sarah came in the evening.

Read two admirable sermons by Kingsley— the

* Afterwards Lady Blackett.—(F. J. B).

1867. tenth and eleventh of his new volume on " Ruth,"
and on "David and Jonathan." That on the story
of Ruth is especially beautiful.

Tuesday, 30th April.

Visit from Mr. Grey and Miss Shirreff, who gave
a sad account of Poor Richard Napier's state. We
heard of the birth of another girl to Harry and
Norah Bruce. Read an article in the new number
of the *Edinburgh Review*, on Louis XV. It is really
a most melancholy history. Though it might be
difficult to find a public character more contemptible
than Louis XV., yet I am disposed to feel even
more pity than contempt for him. He does not
seem to have had originally an evil nature: it was
not at all one of those mysterious cases of natural
depravity. He would probably in any case have
been always a weak character: but his education
was miserable, and the people and circumstances
that surrounded him from his childhood, left him
hardly a chance of good.

Wednesday, 1st May.

I finished reading a little book on the "French
Revolution," by H. Carnot (son I believe of the
great Carnot) sent to me from Paris by Mrs. Byrne.
It professes to be a "Résumé historique," and is
in fact a very rapid, lively animated sketch
of the History of the Revolution during the
two first assemblies (to September 1792) from
a thoroughly republican and democratic point
of view. It is thoroughly one-sided, the author

cannot see a particle of good in a king, a queen, 1867.
or an aristocrat; but I must own that he does not
make any attempt to palliate or excuse the massacres
of September.

By Sir Edmund Head's recommendation, we
went to Colnaghi's to see a new engraving (by
Robinson (?) after Vandyck's portrait of Anne Carr,
Countess of Bedford, at Petworth; and we ordered
a proof. It is a beautiful engraving.—Sir Edmund
Head says it is the finest that has been executed in
our time, and he does not think it likely that any
thing like it will ever be done in future. We then
went to Lambert's* and to Howard's† and then
called on Mrs. George Jones‡ and Mr. Eddis.

I went to Bentham's reception (as President of
the Linnean Society) at the Linnean Society's
rooms at Burlington House. The rooms were very
beautifully decorated with a profusion of fine living
plants in flower, from Kew and from various nursery
gardens,—many of them rare and curious, especially
the very singular Orchid, *Uropedium Lindenii*. On
the tables were exhibited a great many curious
and interesting objects of natural history: rare birds
and insects brought by Mr. Wallace, from the
Eastern Archipelago; a collection of various cones
of the Coniferous and Proteaceous orders, curious
fruits of Bignoniaceæ, and some other families from
Brazil: and above all, fine specimens of that most
wonderful and beautiful Sponge, Euplectella
Aspergillum. On the walls were a variety of
prints, drawings and photographs.

* The Jeweller.   † The Upholsterer.   ‡ The Widow of the Artist.

Minnie and Sarah and Miss Kinloch came to
luncheon with us. We went to the private view of
the National Portrait Exhibition, at the South
Kensington Msseum. It is a rich and very interesting exhibition. The Hogarths, Gainsboroughs and
Reynoldses well arranged. I had sent five pictures
from Barton: — four Reynoldses — Mrs. Bunbury,
Mrs. Gwyn, Mr. Bunbury (the picture which I had
bought two years ago) and Sir Thomas Charles
Bunbury, and one Kneller—Sir Thomas Hanmer.

The portraits which I particularly noticed, in this
and subsequent visits, were—Lady Coventry and
the Duchess of Hamilton by *Henry Morland* (most
fascinating). Mrs. Sheridan as "St. Cecilia," by
Reynolds (the Bowood picture, exquisitely beautiful).
"Miss Linley," (the same Mrs. Sheridan) and her
brother by *Gainsborough*. Lady Waldegrave
(Duchess of Gloucester)—*Reynolds*. Mrs. Horton
(Duchess of Cumberland)—*Gainsborough*. Mrs.
Graham, by *Gainsborough*. Nancy Parsons—*Gains-
borough*. All these ladies are lovely. Lord Lovat—
Hogarth. Lord Chesterfield (at age of 76)—*Gains-
borough*. Miss Rich—*Hogarth*. Two caricatures—
Reynolds. Oliver Goldsmith—*Hogarth*. Sterne—
Reynolds (from Bowood). Baretti —*Reynolds*. The
Holland House picture (celebrated by Macaulay).

A beautiful day—quite summer. We went to the
private view of the Royal Academy. A very good
exhibition. I particularly noticed *Millais'* "Jephtha,"

his two truly charming pictures of a child entitled 1867.
"Sleeping" and "Waking." *Cooke's* "Canal of
Giudecca, Venice." *Frith's* "Charles the Second's
last Sunday." *Hook's* "Digging for Sand-eels." *Creswick's* "Beck in the North Country." *George Jones's*
"Malines." *Linnell's* "Harvest Showers." *Eddis's*
"Bird's Nesting." *Poynter's* "Israel in Egypt," and
Leighton's "Venus."

[NOTE.—Of *Leighton's* "Venus," my opinion has
changed on a subsequent view since I have been at
Paris. The figure and attitude are graceful, but the
colouring appears to me lifeless and unnatural:
ivory rather than flesh. Still I like it better than
any of the other pictures here by the same artist.
Poynters "Israel in Egypt" appears to me one of the
finest historical pictures that I have seen for a long
time:—wonderfully real and animated. It gives one
a more forcible idea of the amount of tyranny on the
one side, and of degradation and misery on the other,
that went to the building of those wonderful works
of Egypt than anything I ever read about them].

Met at the Royal Academy, which we saw very
pleasantly, Mrs. Tait, Lady Colebrooke and Miss
Richardson and others.

Visit from Mary and Katharine in the afternoon.

Saturday, 4th May.

A long talk with Scott—just come to town about
Mildenhall business. I called on Mr. Robert Scott
at the Meteorological office in Parliament Street,
to see a collection of fossil plants from Greenland,

1867   which he has in his care; they belong, as I understand, to the University of Dublin, and they are the
specimens which Heer has described and figured
in his forthcoming work on the Fossil Flora of
the Arctic regions.   They are very remarkable,
especially as proving beyond a doubt that there was
a time when that dreary country was covered with a
rich forest vegetation, probably very like that which
now flourishes in British Columbia.   I went to an
evening party at Lady Colebrooke's.

Monday 6th.

A beautiful day and very hot.   I wrote out a
sketch of a botanical essay for Kingsley.

We went out in the open carriage, picked up
William Napier at his office, took him to Harley
Street, and then back through Hyde Park to Sloane
Street.

We went to the Portrait Exhibition, and then to
Battersea Park.

Tuesday, 7th May.

I looked into the Royal Academy, and went to
the Horticultural Society's Exhibition (which I
visited but hastily), saw splendid Azaleas and Roses
and a great variety of very beautiful and curious
Orchids: but not many that I had not seen before,
either at Veitch's or at Mr. Bentham's Soirée.   No
striking novelty in the Ferns.   One very striking
Epidendrum (I think) with such tall slender arching
stems, bearing regular distichous leaves towards
their tops, that at first sight I took them for some

kind of dwarf Palm. Here, as at Mr. Bentham's, 1867. there were several fine specimens of a very curious and handsome Aroideous plant, Anthurium Scherzerianum with beautiful bright scarlet spathe and spadix. Also several varieties (as I suspect rather than species) of Bertolonia with variously spotted and variegated leaves and a very curious plant "Dalechampia rosea" (from Mexico I think) with large involucral bracts of a delicate rose-colour, reminding one at first of Bougainvillea, but it is really next of kin to Euphorbia.

Our dinner party—Mr. and Mrs. Bentham, Mr. and Mrs. Lombe, and Julia, Minnie, Katharine, Mr. Eddis and Charles Chichester. Sarah in the evening.

Wednesday, 8th May.

Weather beautiful. Made good progress with my botanical essay for Kingsley. Read Kingsley's noble sermon on "Faith." We visited the Charles Lyells, who are going to Lynn for a fortnight. Geological Society meeting.

Thursday, 9th May.

Visit from Mrs. Mills.

Saturday, 11th May.

Minnie and Sarah, Sarah Hervey, Douglas Galton, and Edward dined with us; and all the party, except Fanny, went to Murchison's geographical party.

Here is a long gap. I must recapitulate. On the
11th May, exactly this day month, we received the
news of the almost sudden death of Fanny's Aunt,
Mrs. Power, at Paris. We immediately determined
to go to Paris, that Fanny, as all her sisters were
absent, or otherwise engaged, might be able to help
and comfort in some degree her surviving Aunt,
Mrs. Byrne. We arrived in Paris on the 18th, and
remained till the 29th. It was not the season I
should have chosen for visiting Paris, as the great
city was ten times more than usual the head-
quarters of bustle, hurry, noise and excitement. It
was a huge fair, in which almost everybody seemed
to be intoxicated with excitement of one kind or
another. I spent two long mornings at the Exposition, besides a shorter visit with Fanny:—of
course not half enough nor a tenth part enough
to see it thoroughly ; yet pretty nearly enough
to satisfy me. It is no doubt a wonderful collection
of almost everything imaginable, and I suppose
is in reality richer in various departments than the
last London Exhibition, but of this only *experts* can
judge, and it is certainly very inferior in general
beauty, even to the Exhibition of '62, *much more* to
that of '51. In fact it has *no* beauty of general
effect, there is no *coup d'œil*. On the other hand,
the arrangement is certainly admirable, and there is
a great deal which is beautiful as well as entertaining, even in a hasty view. The porcelain, the
glass, and the tapestry are very beautiful, though
I do not feel qualified to judge whether they are

finer than those exhibited in '62. In the Fine Arts 1867. department there is an extensive and fine display of French pictures, well calculated I should think to give one a good idea of the peculiarities and merits of all their best modern artists, and it is interesting to compare the performances in this way of the different European nations. The British school, however, is rather poorly represented. The Italians appear especially strong in sculpture, and above all there is a noble statue of the first Napoleon in a dying state, which was a great object of general admiration.

There is an interesting collection of antique, or rather mediæval works of art : and what especially struck me as curious and remarkable, a singularly rich collection of works of the "Prehistoric" times, flint implements, &c., and above all a great variety of articles of the Reindeer period from the caves of Doraogne. Joseph Hooker, whom I have seen since my return, had likewise been especially struck by the collection of "Prehistoric" antiquities, and thought it singularly curious and valuable : he does not think there is anywhere another like it.

In another part—in one of the "rues" of the French department, my attention was caught by a most magnificent specimen of emeralds in their natural state, by far the finest I have ever seen.

The "Great Exhibition" was by no means the only exhibition open in Paris whilst we were there. I went twice to the "Salon," (the exhibition of pictures of the year corresponding to our Royal Academy exhibitions), and repeatedly to the glorious

1867. galleries of the Louvre. The pictures at the Salon were very numerous, much more so than I ever saw at any exhibition in London, and the general effects very brilliant and striking. I was struck in general with the boldness of the French painters, both in the choice and treatment of subjects as compared with the English; and this may sometimes be carried too far. The chief fault I find with the French is their love of horrors, their taste for painful or terrible subjects. Pictures of shipwrecks, conflagrations, massacres, tortures, and a variety of horrors, are conspicuously numerous.

There is actually a picture of a wretched man undergoing the torture; not a martyrdom but the torturing of a criminal; and another, an enormous picture, of a burning of heretics in the presence of the Duke of Alva. I think it positively atrocious to exhibit such scenes.

Battle pieces which abound in the "Great Exhibition" are few in the "Salon," but there is one very striking and painful picture of a battle field after the fighting is over; the ground strewn with dead and wounded men and horses, a group of riderless horses gathered together on a rising ground, one of them neighing, and one of the poor wounded creatures stretching its neck and neighing in reply, with a wonderful force of expression. Pictures of naked women which are now so rare in our exhibitions are very numerous here.

On the 24th of May (Friday) we dined at the Tourgueneff's and met Madame de Peyronnet and her two daughters—very nice girls; also Mr. Chad-

wick. Altogether a very agreeable evening. Madame 1867
de Tourgueneff and her daughters charming.

We travelled from Paris on Wednesday the 29th
with Sir John Kennaway and his daughter and
sister. We had a very good passage.

Spent the evening at Folkestone with the
Kennaways, and travelled with them to London on
Thursday, the 30th May. The 23rd anniversary of
our wedding. Thanks be to God.

The Kennaways were very pleasant.

We arrived safe and well at 48, Eaton Place—
Thank God.

May 31st.

A visit from that charming girl, Rose Kingsley.

Called on Mary Lyell. Found Katharine with
her, and had a good talk with them.

Monday, 3rd June.

A dinner party at home. Lady Head, Sir John
and Miss Boileau, Joanna Richardson, Mrs. Mills,
William and George Napier.

Tuesday, 4th June.

We heard of the arrival of Susan and Joanna
Horner from Florence, in Harley street, and Fanny
went to see them. I visited the British Institution.

Tuesday, 11th June.

Went with Fanny and Susan to Colnaghi's ; we
settled to buy a portrait of Gaston of Orleans.*

* Believed to be by Gonzalez Coques.—(F. J. B.)

Wednesday, 12th June.

Sir John and Lady Hanmer came to luncheon with us.

Fanny was obliged to go alone to dine with Mrs. Berners to meet Lady Anstruther.

Friday, 14th June.

Theresa Boileau, Sir John Kennaway and Clement came to luncheon with us.   We dined with the Adairs—a large party.

15th June.

Very cold weather.

A visit from Lord Stradbroke.   Norah Bruce and Katty Napier came to luncheon.

Fanny went to a party at Miss Coutts's.

Before the end of April I had received a letter from Kingsley, telling me that he was temporary editor of *Fraser's Magazine* in the absence of Froude from England, and begging me to contribute to it an article on botany.   I consented, and began to write an essay on some of the geographical relations of the vegetation of South America, containing some ideas which had been a good while in my head. I had made some progress with it before we were called to Paris;   I resumed it after our return, finished it on the 7th of this month and sent it to him, and on the 10th I had a letter of approval and thanks from Kingsley.   It is to appear in either the July or August number of *Fraser*.

Since finishing this I have resumed my Biographical Memoirs of my Father.

While we were at Paris, I began reading Jesse's " Life and reign of George III." and have gone on with it since. It is very pleasant reading : not containing much that is new or very striking but well put together, written in a clear and easy style and in a good spirit.

I have also read some of Froude's essays lately published under the title of "Short Studies on Great Subjects ;" that on the "Times of Erasmus and Luther," is particularly interesting.

Monday, June 17th.

Our dinner party consisting of the Adairs, Thornhills, Charles Lyells, William Napiers, Mrs. Berners, Minnie, John Strutt, Edward, Douglas Galton, Arthur Milman. A very pleasant party.

I have omitted to mention that last Monday (10th) we visited Mr. Cooke, the celebrated painter of sea-coast views, at his house at Kensington Gore and saw the last pictures he has painted of Venice —very beautiful. His house is a perfect museum, *crammed* with curiosities, both of art and nature, for he is a man of science as well as an artist. His collection of Venetian glass, which seems to be his especial hobby, is wonderful ; and besides a vast variety of artistic and antiquarian varieties, he has minerals, fossils, and rich collection of curious cones and fruits and seeds.

Afterwards we paid a hasty visit to the flower show at the Horticultural Society's gardens. The display of flowers in the great tent was remarkably beautiful, both from the profusion and gorgeous

1867. brilliancy of the flowers, and their arrangement; and the general effect of it was rendered uncommonly picturesque by a number of fine tree-ferns from Veitch's, which were so placed as to form commanding features of the whole.

Tuesday, June 18th.

A very fine day. We went in the barouche to Twickenham to see poor Richard Napier. Found him in as melancholy a state as I had expected, but very kind and friendly to us. Mrs. and Miss Shirreff and Emily Napier were with him. We dined there.

Thursday, 20th.

George Napier came to luncheon. Minnie, Sarah and Edward afterwards, and we all went together to a garden party at General Fox's pretty house in Addison road.

I may put down here one or two anecdotes told me the other day by Charles Lyell, and which I omitted to note :—

After Lord Derby had published his translation of Homer, some one said to him that he had been surprised to find that Lord Stanley had not read anything of his Father's Homer. "He"! replied Lord Derby, "he will never read it till it is turned into prose and put into a blue book!"

At a dinner party at Sir J. Packington's, Lord Derby and his son were both present, and soon after the ladies left the room, Lord Stanley got up and in spite of the remonstrance of his host, went 1867. away to his business. As the rest of the party were settling themselves again after his departure, Lord Derby said, "Now that the old gentleman is gone, let us enjoy ourselves!"

Shafto Adair dined with us on the 17th. I was sorry to hear him say that he thought it a mistake in the Government to spare the lives of the Fenian leaders — Burke and McCafferty; and that the middle class people as well as the gentry in his part of Ireland were of the same opinion. In spite of this I am not convinced that he is right.

What Shafto Adair said about the cause of Fenianism was quite in accordance with the conclusions I had already arrived at. He says (exactly as I had judged) that Fenianism (or rather perhaps the general feeling of disaffection which leads the lower classes of Irish to favour the Fenians) is in its origin entirely dependant on questions of territorial right. The meaning of it is that the Irish peasantry believe that they have an absolute right to the land. As Adair stated it—every estate and every farm in Ireland has two owners—the owner in possession and one who believes himself to have an indefeasible ancestral claim. This is perhaps a more terrible aspect of Irish discontent than any other.

Katharine and Leonard dined with us.

Friday, June 21st.

Went to the Linnean Society. Fanny spent the day with Emily Napier, at Wimbledon. I went

1867. out thither to dinner. Met Edward there and Mrs. and Miss Holland.

A fine day.

A visit from Sir Edmund Head, who approved of our new pictures. We went with Minnie to Miss Coutts's garden party at Holly Lodge, a beautiful scene. Afterwards to a party at Katharine's.

Clement arrived.

Our dinner party consisted of Lady Bell, Mr. and Mrs. Hutchings, Mr. and Mrs. Godfrey Lushington, Pamela and Philip Miles, Minnie and John Herbert, George and William Napier, Edward, Mrs. Young and Sally.

Transacted business with Mr. Weight.

Visit from Admiral Eden. I went to the Portrait Gallery with Minnie and Sarah. We went leisurely and quietly through nearly the whole and saw it very pleasantly. Met Sir Frederick Grey there.

Visited Kew Gardens with Katharine and we spent more than three hours there very pleasantly ; we also visited the Hookers.

The gardens are in great beauty, although many of the more tender exotic Conifers and some other

trees and shrubs have suffered from the last winter. 1867.
The Palms, Musaceæ, Pandani and Bamboos in
the great Palm-house are magnificent; but the
grand Ensete is dead.

I admired the museum of specimens of woods and
sections of trunks which I had not seen before. A
vast variety of beautiful and richly coloured woods
from Australia, India and South America; planks of
Deodara of vast size. Sections of huge trunks of
White Oak from North America, &c., &c.

The collection of tropical Ferns is amazingly
rich and beautiful, in flourishing health, and carefully labelled.

I was surprised at the number of species of
Trichomanes and Hymenophyllum, which are here
growing vigourously and beautifully in glass cases.
A few years ago it was thought almost impossible to
cultivate them. Leptopteris superba and pellucida
in the same situation are not as fine as at Veitch's.
An excellent plan here adopted in the tropical Fern-
house—pieces ("tronçons") of the trunks of Tree
Ferns are set up, and small and delicate Ferns of
various kinds are planted on them, where they
flourish exceedingly, just as one sees them on the
same tree-ferns stems in the tropical forests.

In a smaller tropical house, I admired several
flourishing plants of Cephalotus follicularis with its
exquisitely beautiful and curious pitchers; also two
or three species of Saracenia and several Droseras,
especially the very curious Drosera binata; all
these growing in pots filled with sphagnum.

In the great cool conservatory, some fine Arau-

1867. carias, Eucalyptuses and tropical Acacias, but as yet the general effect of the vegetation is not as fine as in the Palm-house. Nor do the Proteaceæ strike me as being by any means as fine as what I saw many years ago in the Botanic Garden in Edinburgh. Here is a fine new Araucaria, A. Saviana, (named and described by Professor Parlatore) from Bolivia nearly allied to A. Brasiliana, but with much narrower and more widely-spreading leaves. Dr. Hooker seeemed to think it might prove to be a variety of A. Brasiliana, but it appeared to me (as far as the leaves go) to be more different from Brasiliana than *that* is from imbricata.

Also Rhododendron Maddeni, one of the Himalayan species with magnificent white blossoms, deliciously sweet scented, almost equal to Gardenia. Also Maclenia speciosissima from New Granada, a kind of glorified Vaccinnium with a profusion of large waxy, bright, orange-coloured flowers. It has quite the habit of Thibandia. How does the genus differ ?

June 27th.

Palmer set off for Barton with a good deal of luggage.

We went with Minnie and Sarah to see Mr. Holford's splendid house and pictures—afterwards to several shops. We dined with the Rayleighs—a large party—many more in the evening.

June 28th.

A talk on business with Mr. Nichol. A farewell

visit from dear Minnie and Sarah. We dined with
the Thornhills. I sat by Mrs. Angerstein, and
found her very pleasing. Afterwards we went to a
large evening party at Mary's.

June 29th.

A beautiful day. We had a very pleasant journey
down to Barton in our own carriage with post
horses ; did it in nine hours and a half. Arrived
safe at home, and found all well, thank God.

July 1st.

Beautiful weather. Busy unpacking, arranging
and settling ourselves at home, Walked about the
grounds, inspected the young trees, ferns, &c.; went
in the pony carriage with Fanny, to call on Mrs.
Rickards.

2nd July.

Very warm. A heavy thunderstorm in the afternoon.

Transacted business. Strolled through the
gardens and grounds, enjoying the beauty of the
season.

LETTER.

Barton,
July 7th, 1867.

My Dear Mary,

We have had a most comfortable quiet
week, and now are screwing ourselves up for our ten

1867. days of bustle.   The weather here is rather cold,
but the roses are in all their glory, the hay-making
going on famously, and the place altogether most
enjoyable.

Ever your affectionate Brother,

CHARLES J. F. BUNBURY.

JOURNAL.

Monday, 8th July.

Sarah and Kate Hervey came and spent the
afternoon with us.—Sarah charming.   Visits also
from Lady Cullum and Captain and Mrs. Horton.

9th July.

Glorious weather.   I was much out-of-doors.   I
saw a white owl flying in the full sunshine.
Received a pleasant note from Kingsley.

10th July.

Went on with my catalogue of South Brazilian
plants.   We dined with the Wilsons at Stowlangtoft.   Met Sir Edward and Lady Caroline
Kerrison, Mrs. Rickards, &c.

11th July.

Mrs. Adair, Sir Frederick Grey, Edward and
Henry arrived.

12th July.

The Wilsons and Sir Edward and Lady Caroline
Kerrison came to luncheon with us.

We all went to the Agricultural Show at Bury.* 1867.
Spent an hour-and-a-half there, — made some
purchases. Lady Cullum and Mr. and Mrs. Hall
Dare dined with us. Sir Frederick Grey very
agreeable.

———

13th July.

A great thunderstorm. Shafto Adair and Mr.
George Lock arrived.

———

Monday, 15th July.

Excessively wet weather. Most unfortunate for
the Agricultural Show.

Mr. and Lady Mary Phipps came to luncheon
with us. Lord Stradbroke, the Anstruthers, Mr.
Boileau and Mr. Lombe arrived. The Arthur
Herveys dined with us.

Read chapter 22 of Burton:—the capture and
execution of Wallace: Edward's attempt at the
pacification of Scotland: Bruce's escape from the
English court, and his insurrection: the murder of
Comyn: the coronation of Bruce, and the death of
Edward I. Burton is very fair and impartial in
his narrative of these affairs, though he naturally
and most rightfully sides with the Scots, he does
justice to the great and statesman-like aims of
Edward. He thinks that in the interval between
the destruction of Wallace and the escape of Bruce,
that is about the latter part of 1305, and the
beginning of 1306, when Scotland was apparently
subdued and pacified, the intentions of Edward, as

* The Royal Agricultural Show took place that year at Bury.—(F. J. B.)

1867. indicated by his ordinances, were really those of a great and benevolent though despotic ruler; that his object was to fuse England and Scotland into one compact and thoroughly united kingdom, but the hatred which his previous measures had kindled, made the accomplishment of this purpose impossible.

16th July.

We all went into Bury in the morning, and spent about two hours at the Agricultural Show. Met many acquaintances. Sir Frederick Grey went away early. Captain Horton came to luncheon. Mr. Gurdon arrived. Mr. Borton dined with us.

Wednesday, 17th.

The Adairs and Lord Stradbroke went away. Lady Cullum, Lord and Lady Wm. Graham and the Milner Gibsons came to luncheon. We went to the Horticultural Show at Bury, and were caught in a storm of rain coming away.

Thursday, 18th.

Henry went away *early*,—the Anstruthers, Mr. Boileau and Mr. Gurdon after breakfast.

This was my dear wife's 53rd birthday. We dined with Lady Cullum.

LETTER.

Barton,
July, 18th, '67.

My Dear Katharine,   1867.

Many thanks for your letter. I was very glad to hear of your being again safe on this side of the Channel, though very sorry you had such a bad passage; but I trust you will not be seriously the worse for it. I am sure you must have been much in need of rest. The weather all this week has indeed been very rough, and disagreeable, and must have given our foreign visitors a dismal idea of our climate. Monday was a true St. Swithin's day, and the unfortunate Viceroy saw Bury and the Agricultural Show under a rain which seemed to intend to wash everybody away. He must have thought it very unlike Egypt. Sir Frederick Grey said that the mud in the Show-yard reminded him of Balaclava. But the ground hereabouts dries quickly, and when we were there on Tuesday we could only judge of what the mud *had been*. The bad weather, I am afraid, must have seriously interfered with the receipts in money, but all whom I have spoken with agree that the show has been an extremely good one, both in implements and animals. The Horticultural Show, which is held in the old vineyard of the Abbey, opposite to the Botanic Garden, is very pretty; not equal it strikes me, in the plants exhibited, to some I have seen in London, but there are many beautiful things. I

1867. am very glad that Allan has got one of the highest prizes for Ferns.

Our party is now nearly broken up: only Mr. Lombe and Edward remain till tomorrow, and will go with us to dine to-day at Hardwick. We have had several pleasant people, especially Sir Frederick Grey, who is one of the most delightful men I know.

Fanny is well, but I think that when all our guests are gone she will feel the fatigue: for this party has given her much more than usual trouble, and has tried to the utmost her powers of organization: but she is equal to any emergency.

With much love to your husband and children,

I am ever,

Your very affectionate Brother,

CHARLES J. F. BUNBURY.

JOURNAL.

Friday, 19th.

Mr. Lombe and Edward went away, and we were again alone. Wet, stormy, dismal weather.

Sir Frederick Grey gave us an interesting account of an expedition he made with Lady Grey up one of the mountains in the Isle of Bourbon (or "Reunion") where they had to help themselves up the almost precipitous ascents by the roots of trees, and were attended by a splendid official guide with a great cocked hat and sword, and he spoke with enthusiasm of the magnificent tropical vegetation,

the Ferns and Palms on those mountains. He told 1867.
us also of the great Chinese river, the Yang-tse
kiang, which he had ascended in a ship of war as
far as Nankin.

Off the mouth of the river, he said, the muddy
sediment is apparent—out of sight of land. He
believes that a great Delta, like that of the Ganges
will be formed there.

Sir Frederick Grey is a most happy combination
of the spirit and vivacity of a sailor with high cultivation and perfect refinement. He has been in almost
all parts of the world, at any rate in all the principal
divisions, and has everywhere observed much, not
only with natural and professional acuteness, but
also with that accuracy which is given by the study
of the natural sciences.

LETTER.

1867.   My Dear Lyell,

I owe you a great many thanks for your kind letter of the 9th, which gratified me much. I am exceedingly glad that you like my botanical article in *Fraser*; and you have pointed out just those parts of it with which I am myself best satisfied. With respect to the Darwinian "moral" which you wish I had "pointed" more distinctly:—I was apprehensive that my essay might be too long, and besides, I am not prepared to go quite so far in Darwinianism as you do. Perhaps my complete conversion is reserved for your second volume.

I think there are points in my *Fraser* essay, about the distribution of forms as influenced by the distribution of mountains and plains, which might deserve to be more fully developed, with more of scientific detail and illustration. In fact, I had long been thinking of writing something on the subject for the Linnean Society, but had not got together sufficient details for a thorough scientific paper; and Kingsley's request just gave me the opportunity of throwing off a first sketch of the subject.

Have you heard from Kingsley? He told me he wanted to write to you, to ask your opinion as to the Bagshot Sands in reference to the Glacial Period. He says he has collected facts which leave no doubt in his mind, that the whole of the Bagshot

sand district was under the sea, and traversed by 1867.
icebergs in that period, and that during that period
it rose slowly out of the sea. He asked me to
mention the subject to you.

I hope you and dear Mary are quite well, and will
enjoy your visit to Paris. We are now again quite
alone, and not a bit the worse for the excitement of
the Agricultural and Horticultural Shows at Bury.
We had a very pleasant party on the whole, though
with too great preponderance of the masculine
element; and in spite of the wretched weather, I
believe both shows were considered very successful;
certainly they gave great pleasure to the neighbourhood, and I have no doubt will have been very
useful.

I tried to get Mr. Berkeley ("the Fungus man," as
Fanny calls him), the editor of the *Horticultural
Society Journal*, to come to us for the show: but he
was engaged to an examination at the London
University.

My gardener got prizes both for British and
exotic Ferns, and also for Gloxinias and Achimenes,
and I believe he could have got one for Roses if he
had tried. Certainly the Roses have been most
lovely ever since we came home, and are so still in
spite of the rain.

Much love to dear Mary,

Ever yours affectionately,

CHARLES J. F. BUNBURY.

Almost the only plant we have lost by the severity
of this last winter or spring, is Magnolia glauca,
the most northern of all the Magnolias The

1867. Halesia (Snowdrop tree) and the Æsculus macrostachya, both of which are natives of Carolina, are as flourishing as possible. It is very odd.

JOURNAL.

Monday, 22nd

We took a pleasant drive to Livermere and Ampton, and called on Mrs. Horton and Mrs. Rodwell. The Agricultural and Horticultural Show at Bury, which ended last Friday, 19th, were in spite of the bad weather very successful on the whole: and very great treats to the people of this neighbourhood, and I have no doubt very useful. The Horticultural Show (which I understand better than the other) was certainly very good:—not so rich perhaps in some departments, especially Orchids, as some I have seen in London, but very beautiful nevertheless. Splendid yellow Allamandas and rose-coloured Dipladenias: gorgeous scarlet and orange-coloured Ixoras, and the finest specimens I ever saw of Saracenia purpurea and flava, perfect masses of them. The greatest novelty to me was an extraordinary plant exhibited under the name of Lilium Giganteum from Lord Walsingham's, in Norfolk.

Saturday 27th.

We went to a musical garden party at the James Bevans. Met many acquaintances. Afterwards drove to Hardwicke and had an *extempore* dinner alone with Lady Cullum.

Wednesday, 31st July 1867.

Finished reading "Rokeby" to Fanny. In the course of this month I also ran through Wright's "England under the House of Hanover," an entertaining work in which the history is illustrated by copies from the caricatures and extracts from the satirical writings of the times.

LETTER.

August 1st, 1867.

My Dear Katharine,

I have lately finished Miss Edgworth s Memoirs, and I thank you very much indeed for lending them to us. They have both entertained and interested me extremely. Her character as it comes out in those letters is quite delightful, and I think her genius and wit appear to still more advantage in her letters than in her published works, because she is not always intent on a moral lesson. This is what to me rather spoils nearly all her novels (I do not speak of her writings for children, in which the moral preaching is more in place)—that the moral purpose is too conspicuous, is thrust forward too studiously and ostentatiously. I wanted to read again her "Helen," which (as I remember) I thought the best of all her novels, but which is not included in the collected edition of her works ; Willis & Sotheran told us it was quite out of print, and could only get for us with some difficulty, a second hand (or much more than *second* hand)

U

1867 copy, so dirty that we are obliged to get it new bound before we can read it.

Miss Edgeworth's account of her Connemara expedition is most curious and entertaining. I should like much to know what became afterwards of that Miss Martin, who must have been a very remarkable character. The Editress's part seems to me to be done with much good taste and judgement.

I am reading Burton's "History of Scotland," and have come to the battle of Otterburn, "where the dead Douglas won the field." Burton is a very clear, very careful and very sensible writer, without a particle of poetry or enthusiasm ; the very reverse of Froude. He seems to revel in old records, old Acts of Parliament, legal antiquities ; in fact, I should think he must be a good deal like Mr. Jonathan Oldbuck of Monkbarns.

I congratulate Leonard on his speech. To me, at his age, it would have been a dreadful undertaking. I do not wonder he enjoyed his visit to Pendock; it is an interesting country in many ways.

The Fern which you enclosed in your letter to Fanny, does not appear to me to differ in anything from Filix-foemina.

We saw a good collection of out-door Ferns the other day at Henry Blake's at Hesset, a few miles from here ; they have 33 species and varieties, all very flourishing, and some of them rare, especially Nephrodium (Lastrea) cristatum, which grows finely ; it is a very rare British Fern.

The weather is worthy of Terra del Fuego; we have fires every evening, and I am at this moment writing with a fire in my sitting room, and not at all too warm. I wonder whether it is as cold with you in Scotland; it can hardly be colder.

Pray give my kind remembrances to your Brother-in-law and the Miss Lyells. I can never forget their kindness to us when Fanny was so ill at Kinnordy.

Much love to Harry and your children,

Believe me ever your very affection Brother,

CHARLES J. F. BUNBURY.

JOURNAL.

Friday, 2nd August.

Went on with the Barton Estate accounts, and with my Memoirs. Finished chapter 26 of Burton's "Scotland," comprising the reigns of Robert II. and Robert III., the first two kings of the house of Stuart (from 1370 to 1406). This period includes the battles of Otterburn (1388) and Homildon (1402) the singular incident of the fight between two select bodies of Highlanders in the presence of the King and Court at Perth (1396), and the suspicious death or murder of the Duke of Rothesay (1402).

Examined flower of Locheria.

Mr. Bowyer and Mr. Lott arrived. Our dinner party was Lady Cullum, the Wilsons, Sir Richard Kindersley, Miss Milbank and Mr. and Mrs. Morgan.

Saturday, 3rd.

Mr. Lott went away after luncheon.   Mr. Bowyer
went to Bury in the morning, to Hardwick, and
afterwards returned here to dinner.   We spent
the afternoon most agreeably with the Arthur
Herveys, quite alone with them.   They are
delightful.

<hr />

Monday, 5th.

A beautiful morning.   Discovered the first cones
on one of our Cedars.   Finished the Barton Estate
accounts of last quarter.   Went on with the
Memoirs of my father.   Arranged some dried plants.
We took a drive with Mr. Bowyer.

<hr />

Tuesday· 6th August.

Very wet.   Mr. Bowyer went away.

<hr />

Tuesday, 13th.

Glorious weather.   We drove to Ashfield and
called on Mr. and Lady Susan Milbank and saw
them.

<hr />

Thursday, August 15th.

Finished chapter 27 Burton's "Scotland," comprising the reign of James I.   Burton attaches
much importance to the battle of Harlaw (1411) as
a decisive trial of strength between the Highlanders
and Lowlanders between the Celtic and the
Germanic races of Scotland.   Before that he seems
to think the subjection of the Lord of the Isles
and the Celtic population to the King of Scot-

land was scarcely more real than that of Scotland 1867.
to England a century before, and there was no
evident reason why the Lord of the Isles might
not have established a really independent kingdom,
but the higher civilization triumphed.

James I. of Scotland seems to have been a really
great man; or rather to have had in him the elements
of a great man; he attempted great and enlightened
reforms, but he was too much in advance of his age
and country. He was perhaps harsh and stern in
enforcing his reforms; and he fell a victim to the
resentments which he had thus provoked.

We went through the north grove with Scott, and
marked many trees to be taken out.

LETTER.

Barton,
August 15th, 1867.

My Dear Katharine,

I am delighted to hear that you had such a
good botanizing at Clova. I was going to say that
I wish I had been with you; but that would look
like discontent, which I do not feel, for I was very
agreeably occupied during that same time; but I do
wish and hope I may one day be with you in a
good botanical expedition.

I think one of the greatest pleasures in life is a
good day's botanizing in a fine wild country, and in
pleasant company. I do not know whether I shall

1867. ever enjoy that pleasure again. But I can thoroughly sympathise with and rejoice in the happiness of young naturalists like your Leonard. I cannot I am afraid, quite so well enter into the delights of fly-fishing, but if I cannot entirely *feel with* Leonard and Frank in this respect, I am very glad they were so happy and so successful.

I am very much obliged to you for your offer of a plant of Woodsia, and shall be very glad to try to keep it. Those Arctic plants are often more difficult to keep in our gardens than tropical ones.

Most of our Ferns, indoor and out, are looking extremely well, and the garden, arboretum, &c., in great beauty, and most enjoyable during the splendid weather that we have lately had. The Catalpa tree in the arboretum is one mass of blossoms, quite magnificent. The whole air about here is loaded with the scent of flowers. This last week has been glorious harvest weather, only yesterday the heat was so excessive that the men were absolutely forced to give up their work for a time, and some of them felt really ill from it.

We spent the latter part of last week, Wednesday to Saturday, very agreeably with the Boileaus, who I have a great regard for. Sir John looks, and is I am afraid, sadly broken; but he was as kind and cordial as possible, and was able to take us one day to see the interesting Roman camp at Caistor, and another day into Norwich. The two younger girls are charming and loveable, and there is to me something very interesting in the elder one. We met the Lombes too at dinner there, and spent a couple of hours very pleasantly with them the next day on our way home.

Ever your very affectionate Brother,

CHARLES J. F. BUNBURY.

JOURNAL.

Saturday, 17th August.

We went to Mildenhall and spent the day there, returning to dinner. Saw poor old Mrs. Bucke and Mrs. Marr and Elmer and his wife.

Tuesday, August 20th.

I have lately skimmed Jukes's description of the interior of Java, in the second vol. of "The Voyage of the Fly,"—*most* fascinating. The Highlands of Java must be one of the most delicious countries on the face of the earth.

Saturday, August 24th.

We returned yesterday evening from an extremely pleasant little tour of three days in Norfolk, in company with Arthur Hervey and his two admirable daughters, Sarah and Kate. Sarah in particular is one of the most charming girls I have ever known or can imagine. We enjoyed it exceedingly. The weather has been perfect the whole time, and everything has turned out as well as possible. We (five altogether) started at ten on the 21st, went in

1867. our barouche to Thetford, thence by railway to
Wymondham, Dereham and Walsingham. Fine
remains of the famous Priory of Walsingham, now
standing in the grounds of Mr. Lee Warner's house.
The Shrine of Our Lady of Walsingham was a
famous object of pilgrimages all through the Middle
Ages down to the Reformation. The most striking
part of the ruins, an extremely beautiful arch of vast
height, well represented in Britton's Architectural
Antiquities. The wishing wells—two small circular
reservoirs full of water near the arch, have
superstitions connected with them.

Mr. Lee Warner proved to be a school-fellow of
Arthur Hervey's, and was very courteous and
hospitable. His grounds are beautiful, with many
fine trees, especially a black Walnut, the finest I
have ever seen, except the one planted by Bishop
Compton, at Fulham. Walsingham church has a
remarkably fine font.

We returned to the railway at Fakenham, visiting
on the way a very curious old house (now a farm-
house). At Barsham an uncommonly fine specimen of ornamental brick-work, of the time (I
understand) of Henry VII. We slept at Dereham.
The 22nd was a delightful day. We went, first by
railway to Swaffham and saw the church there, a
large and very handsome one. The tower very
fine—with elaborately wrought battlements at top.
The wooden roof of the church of extraordinary
richness and beauty; thence we went in an open
carriage through a very pleasant varied country to
Castle Acre, where we found plenty to occupy us

very agreeably for several hours. Remains of the 1867.
old Norman Castle very striking and picturesque
indeed, the Keep especially. The massive fragments
of its walls—immensely solid—crown the top of a
high conical grassy mound, excessively steep, which
rises out of an immense ditch. There are great
remains too of the walls inclosing the body of the
fortress and extensive and complicated outworks.
The extent of the place altogether is great, the
ditches and encircling earthworks on a grand scale
and in the middle-age system of warfare it must
have been as strong as a place could be. The
ground slopes down from it into the valley through
which the pretty little river Nar winds. Sheep
feed on the grassy slopes of the ditches and earthworks; it is a pleasant, peaceful, thoroughly rural
scene on which we looked from the old defences of
the Castle. We eat our luncheon very merrily
beneath the old walls of the Keep at the top of the
exceedingly steep mound where we could hardly
find secure sitting room. This Castle was built
very soon after the Conquest by the same Earl
Warrenne, who founded the Priory. Castle Acre
Church is large and well situated, and has been
handsome but is much neglected. Castle Acre
Priory is down in the valley. The Priory was
founded in 1085 for Monks of the order of Cluny,
by the 1st Earl Warrenne, to whom the Conqueror
granted not less than 140 Manors in Norfolk. The
church was consecrated between 1146 and 1148
(Britton). The remains of the Priory are very
beautiful and very interesting, especially the West

1867 front, which remains in great part perfect, and is a
very striking specimen of the richest and most
highly-ornamented style of Norman architecture.
There are good engravings of it in Britton's
Architectural Antiquities. Sarah Hervey made
some admirable sketches of these ruins, while we
lingered under the Lime trees in the pleasant green
meadows that slope down to them.

I observed Verbascum pulverulentum by the
road-side near Castle Acre, and Calamintha
officinalis (or its variety Nepeta) in great abundance
all over the mounds and ditches.

The 23rd we first visited the Church of Dereham:
it is chiefly remarkable as the burial place of
Cowper. One large monument—excessively plain
—bears tablets in memory of him and of Mrs.
Unwin and of Mrs. Margaret Perowne, who nursed
him in his last illness, with inscriptions in middling
verse. There was nothing either in the monument
or epitaph to engage our attention, but the name
sets our memory and thoughts at work on the
history of one of the most amiable and most unhappy of poets.

Thence we drove to Elsing, about five miles from
Dereham, and saw the church, in which is (in the
floor), a very fine brass of one of the Hastings
family, temp. Ed. III.

In the way back to Dereham, we stopped to look
at the Church of Swanton Morley, another large
and handsome church with a fine tall tower. From
several points of the road, and especially from the
churchyard of Swanton, there are good views of

Byelaugh Hall. Dereham and Swaffham are both 1867. of them very neat, cheerful, thriving looking towns. The King's Arms at Dereham is a very comfortable inn. The country rich, cheerful and very highly cultivated, and at this harvest season very agreeable to look at. Around Dereham, it has very much the same character as our Bury country; about Swaffham it appears more varied in surface, and there are some good bits of open, breezy, furzy land. The river Nar looks like a trout stream.

At Wymondham, where we met Sir John Boileau and his daughters, having to wait for a train, we took the opportunity to see the church, which is a very fine and remarkable one. Fanny and I had seen it twice before while staying at Ketteringham, but it was new to the Herveys ; it is particularly remarkable as having two towers, one at each end; the eastern one belonged to the church of the old Abbey or Priory, is octagonal, very lofty and of great beauty. The arch leading into it from the former nave of the Priory Church is also of great height and beautiful proportions. The western tower is square and likewise very tall and fine. Within, the nave of the church is of a remarkable height, with a carved wooden roof, very rich and beautiful, though not equal to that of Swaffham Church. The north aisle uncommonly broad, with likewise a finely carved wooden roof; the south aisle comparatively very mean—a curious font. In the churchyard are some other considerable remains of the monastic buildings.

We went by railway to Thetford, where the

1867. barouche met us, and on our way home we stopped
at Euston Rectory, Augustus Phipps' where we drank
tea.   Here there is the very finest Horse-Chesnut
tree I have ever seen; some of those in Bushy Park
may be taller, but I have seen none forming so
vast and perfect a dome of leaves, or of so great a
circumference, its branches resting on the ground
for a long way on every side of the trunk; this tree
stands close to the brink of the river, (the little
Ouse) which has probably had some influence in
promoting its growth.   Phipps says it has increased
considerably in size within his memory.

We then drove home, where the Herveys left us;
we were very sorry to part with them.

Lady Louisa Kerr arrived.

_____

Monday, 26th August.

The Miss Richardsons, Sarah Hervey and Mr.
Lott arrived.   We had a very pleasant evening.

_____

Tuesday, 27th August.

I showed Sarah Hervey our Ferns and arboretum.
Drove into Bury with the Miss Richardsons.   The
Arthur Herveys, Lady Cullum and Patrick Blake
dined with us.

_____

Thursday, 29th August.

Our pictures returned from the Portrait Exhibition.
We all went to Ickworth and drank tea with the
Arthur Herveys, and left our dear Sarah Hervey
there.

_____

Saturday, August 31st.

George and William Napier and Clement arrived. We planted trees for Lady Louisa and the Miss Richardsons.

Monday, 2nd September.

Sir Frederick and Lady Grey, Sir Ed. Campbell, Mr. and Mrs. Mills, and two young Lady Legges arrived. The Miss Richardsons sang very agreeably.

Tuesday, 3rd September.

Sarah and John Hervey arrived, and Lady Hoste came to dinner.

Wednesday, 4th September.

Lady Louisa Kerr and the Miss Richardsons went away. I went to the Jail Committee at Bury. Not much business.

Minnie and Theresa Boileau arrived. The Arthur Herveys and Mr. and Lady Susan Millbank dined with us.

September 5th.

Rain in morning, and beautiful afternoon. Went on a little with catalogue of minerals. Gave my Cape Book to Theresa Boileau. Went out driving with Lady Grey and Mr. and Mrs. Mills. We visited Pakenham Church. A merry evening of games.

Friday, 6th September.

Went out driving again with Lady Grey and Mr. and Mrs. Mills, and visited Rougham and Beyton Churches. Captain and Mrs. Horton dined with us; games and much merriment.

———

Saturday, 7th September.

We were all photographed in a group. Then the dear Boileaus and Sarah Hervey went away. The Mills' and their two nieces left us in the middle of the day. I went out in the open carriage with Lady Grey and Fanny. We went to Thurston, Hesset, and Rougham.

———

Sunday, September 8th.

No studies at all this week. The weather beautiful. The house full of company and a party as agreeable in all respects as could be. Sir F. and Lady Grey, Mr. and Mrs. Mills, Sir Edward Campbell, Geo. and Wm. Napier and a charming group of girls. Our dear Sarah, and Kate Hervey, Mary and Theresa Boileau, and two young lady Legges, nieces of Mrs. Mills. It was a delightful party.

Sir F. Grey told us of a nickname given to Sir John Pakington, who has a very large nose; they call him *Marshal Nez !* I hear Sir John Pakington is a man who speaks above himself;—that is, his power of public speaking is out of proportion to his general abilities. This is a cause which in a land of Parliamentary government like this, probably raises many men above their true position.

When the Reform Bill of this year had passed 1867. through the House of Commons, Disraeli being asked by some one how soon it would come on in the Lords, replied—

"You must allow a little time for Derby to make "himself acquainted with the bill. He knows no-"thing of it at present—he knows less than I do, "and that is little enough."

Monday, September 9th.

Sir Fred. and Lady Grey went away. We went, a party of six, to dinner at Hardwick. Met the same party that had been at Barton—a very pleasant party.

Tuesday, 10th.

Sir E. Campbell went away early. Geo. Napier after breakfast and William after luncheon Clement only remaining.

Wednesday 11th.

We went to Wolverstone in our open carriage Our dear Sarah and Kate Hervey with us.

Thursday, 12th.

The party at Wolverstone consisted of Lady Huntingfield and her two daughters, Sir Charles and Lady Rowley and their daughter Edith, young Heathcote Long, Col. and Mrs. Long, and Captain Spencer. Mr. Berners took me through the beautiful fernery on the cliff. We drove to Shotley

1867 and saw the Herveys : they dined with us at Wolverstone.

———

Mr. Berners showed me his garden, hothouses, &c. We all went after luncheon to the Ipswich race course to see the rifle shooting and distribution of prizes. The four Herveys dined at Wolverstone, and all the party, except Mr. Berners, Sir Charles Rowley and I, went to the Ball at Ipswich.

———

We returned home, going a little out of our way to see Wetherden Church, near Haughley.

———

Sally and her daughters arrived.

═══

LETTER.

My Dear Katharine,

I ought to have written long ago to thank you for the Woodsia, which arrived quite safe, and was immediately planted in a pot, covered with a bell glass, and placed in the Fern-house, where it has remained ever since. It has put out several new fronds, and is quite alive, but does not look strong, and I hardly know what is the best way of treating it. Perhaps it is too much *coddled*, and might do better in the cool frame. Considering its native places of growth, it *ought* to be hardy, but so 1867. ought the Parsley Fern, which I have not been able to keep alive either out-of-doors or in the green-house, and which does not thrive even in Mr. Berners's fernery at Woolverstone. I cannot tell whether the Woodsia is Ilvansis or Hyperborea; indeed I do not understand the difference between them.

I daresay you and Leonard and Frank had a very pleasant and interesting time at the British Association.* I did not gather from the accounts in the newspapers which I saw, that there had been any papers of very striking interest or importance, but one really cannot judge from ordinary news-paper reports, and I dare say there was much which you were glad to hear. I should be very glad of a little more information about it. I was pleased at the Dundeen's paying due honour to Charles Lyell.

I think Fanny has sent you from time to time her journals of our proceedings, so that you know all about our delightful three days tour in Norfolk, and our very pleasant party here in the early days of this month. It was a party as agreeable in all respects as could be; one of those which are not only pleasant at the time but leave a permanent impression of pleasure on the mind and memory. Do you not agree with Helps, that the recollection of past pleasure (of course supposing it innocent) is present pleasure? No doubt there is often a tinge of sadness in it, as in my memories of the Cape, associated with so many who are gone, but still, it

* The Meeting of the British Scientific Association at Dundee.—(F .J. B.)

1867. contributes much to the general sum of our happiness in this life.

We spent three days pleasantly enough at Woolverstone. I think I have before described that beautiful place to you, with its fernery on the cliff overlooking the river,—quite a botanical Paradise. Mr. Berners. without being a scientific botanist, is devoted to his fernery and his gardens, so that we are in so far quite congenial spirits: and both he and Mrs. Berners. are exceedingly good-natured and friendly. The rest of the party in the house were all relations or connexions,—Rowleys, Longs and Vannecks; pleasant people enough, too. It was very pleasant taking our dear Sarah and Kate Hervey with us in our carriage to Woolverstone park gate, where their brother met them and took them to his rectory. Sarah Hervey is one of the most charming girls I have ever known.

I have been reading Miss Edgeworth's "Helen," which I had not seen for 32 years; it has not interested me so much as it did the first time, but still I think it is much the most interesting of all her novels,—novels for grown-up people, I mean.

Pray give my very kind regards to your brother and sisters-in-law, and love to your husband and children, and believe me,

Ever your truly affectionate brother,
CHARLES J. F. BUNBURY.

Our great Cephalonian Fir in the arboretum has been measured: it is between 57 and 58 feet high, and 7 feet round the trunk.

JOURNAL.

Kate MacMurdo arrived. Sir Edmund Head 1867. and Mr. Clarke the public orator of Cambridge arrived, also Mr. Lott. We had a very large party at dinner, the Arthur Herveys and their party, seven in number, Lady Cullum, Captain, Mrs. and Miss Horton and Mr. and Mrs. Bain.

———

Sir Edmund Head and Mr. Clarke very agreeable.

———

Sir Edmund Head says that the United States are now undergoing a real, and very serious, though bloodless revolution, of which the ultimate result will probably be, to give supreme and absolute power to the Congress. "State rights" have been already destroyed by the effects of the civil war, and the President will, he thinks, end in becoming the mere Prime Minister of the Congress. The constitution of the United States so much talked of is virtually abolished. The white population of the Southern States now enjoy no more political liberty than the Poles under the Russian Government: and he believes they are so much discontented that in case of a war, they would eagerly join any enemy landing anywhere in the South. Sir Edmund Head remarked, that all confederations from the Achaian

1867. League to the present time, have been formed with a defensive object, to resist dangers from without : and when all external danger is removed, they either break up and separate, or they coalesce and become united under one single government. This last has been the case with Holland, and the United States of America (he believes) are rapidly tending to the same consummation. The Swiss Cantons have been kept together by the continual danger to their independence from France and Germany.

We talked of Homer. Sir Edmund and Mr. Clarke seemed to agree in disbelieving the theory that the "Iliad" is a *conglomerate*. Mr. Clarke said that he had read the "Iliad" in Asia Minor, and the "Odyssey" in the Ionian islands ; and he was strongly impressed with the belief that the two poems were *not* by the same author nor of the same age. The "Odyssey" seemed to belong to a time when the Greeks had somewhat more idea of foreign countries, and much more curiosity about them than in the time of the "Iliad." Sir Edmund said there were several things in the "Voyages of Sinbad" which seemed to be imitated from the "Odyssey."

Sir Edmund Head gave us an anecdote of the Pope. Pius IX. has a great liking for our British representative at Rome, Odo Russel; and one day after praising him, wound up by saying— "Non é Cattolico, ma—io credo che sia pessimo Protestante," as if that were the next best thing.

Mr. Clarke remarked that Juvenal and Martial

are exceptions to the general conventionality of the latter Roman poets; that Martial, with all his faults has a distinct individual character of his own, and does not appear a mere imitator.

But, with the exception of the satirists, most of the Latin poets after the "Augustan age" are very insipid and uninteresting, owing to the want of nature and reality; they wrote on the old subjects of Greek fable, which nobody in their time believed, or even if they chose Roman subjects they endeavoured in the treatment of them to imitate the Greeks as servilely as possibly. This was said with more especial reference to Statius.

Sir Edmund Head said that a passage in Tennyson's "Talking Oak,"—

> " Wherein the gloomy brewer's soul
> Went by me *like a stork.*"

Had puzzled him much; he could not understand why a stork in particular was selected for the simile. At last he found in Brown's "Vulgar Errors," that there was an old belief that storks could not thrive except in a Republic, and therefore they were considered in a manner emblematical of that sort of government. The thought in the poet's mind would therefore seem to be, that the republican spirit typified by the stork was passing away with the soul of Cromwell.

We discussed the expression "Mother-Age" in "Locksley Hall." Sir Edmund Head remarked on its obscurity; Mr. Clarke and I both at once suggested that the author's meaning must be, to personify the general spirit, or the aggregate of the

1867. feelings and tendencies of the present age, and to represent himself (or rather his hero) as the child of this spirit.   Sir Edmund assented.

Wednesday, 25th September.

A beautiful day.

Sir Edmund Head and Mr. Abraham went away. We (with Sally) dined with the Wilsons at Stow-langtoft.   News of the arrest of Garibaldi.

LETTER.

Barton.
September, 26th, 1867.

My Dear Katharine,

Many thanks for your information about the Dundee meeting.   It must have been very pleasant for you.

I have bought Hooker's "Garden Ferns," a beautiful book, and a very good one it seems, as far as I have time yet to study it; for it only arrived yesterday evening.   The more I study my favourite and special botany, the more I feel how infinitely little I know; but I trust to go on learning to the end of my life; and why not in another life too?

I am reading "The Early Years of the Prince Consort."   It is interesting to see his passion for knowledge, even in his early youth, and his strong sense of duty.   I should think it is a book that is likely to be, or ought to be, a valuable lesson to the young generation of gentlemen.

Since I wrote to you on the 19th, we have had a

little more company, Sir Edmund Head and Mr. 1867. Clark of Trinity College, Cambridge (the public orator) ; both remarkably agreeable men. They seemed, too, to suit each other very well. Sir Edmund is really a wonderful man :—such variety and depth of knowledge, such power of expression, such a vigorous intellectual grasp, and with all this, the gaiety and spirits of a boy.

We have had some cold nights lately, but not enough cold to kill down the helitropes or geraniums or to hurt the blossoms of the Cobæa Tritoma Uvaria is just passing off, and Amaryllis Belladonna coming on. Beeches and horse-chesnuts beginning to show a tinge of yellow, and the American Oaks to turn red.

Believe me ever,

Your very affectionate brother,

CHARLES J. F. BUNBURY.

JOURNAL.

Saturday. 28th.

Sally and her daughters went away. Scott brought a very good report of the Mildenhall rent audit. I made progress with my Memoir. Fanny and I dined with the Baines at Flempton. Met the Arthur Herveys, Lady Cullum, and the Bishop of Worcester, and Mrs. Phillpot.

Monday, 30th September.

My Barton rent audit—very satisfactory.

1867.   Wednesday, 2nd October.

Dear Minnie and Sarah and Rose Kingsley
arrived.

———

October 7th.

Sarah and Kate Hervey arrived. They with
Rose and Sarah Napier sang charmingly in the
evening.

———

Tuesday, 8th October.

Mr. Milbank and Lady Susan came to luncheon,
and we showed them the arboretum, School,
etc. I had a pleasant walk with the two Sarahs.
Mr. and Mrs. Henry Hardcastle* dined with us.
She is very pleasing.

———

† Wednesday, 9th October.

I cannot begin this new volume of my notes
without an expression of deep thankfulness to
Almighty God for the innumerable blessings which
He has heaped upon me. I am indeed happy and
prosperous to a degree which I sometimes think is
almost awful. But I may say with Addison—

" Ten thousand, thousand precious gifts
My daily thoughts employ ;
Nor this the least—a cheerful heart
Which tastes these gifts with joy."

We all went in two carriages to Ickworth. Saw
the Arthur Herveys and left there Sarah and Kate
with them. Called at Horringer and saw Mrs.

* Daughter of Sir John.—(F. J. B.)
† Here he begins a new vol. of notes.—(F. J. B.)

Lockhart, and heard a bad report of General 1867.
Simpson.

——— ———

October 10th.

I showed the fern-houses and the arboretum to
charming Rose Kingsley. Took a walk with Sarah.
We met Rose and Kate riding. The Abrahams
and Lady Cullum came to dinner.

———

Monday, 14th.

Mrs. Young and two daughters, and William
Napier with Cissy and Susie arrived.

———

Tuesday, October, 15th.

Emily Napier, Sir George Young, two Anstruthers
and young Long arrived.

———

Wednesday, 16th.

Our dance. Very pleasant, and kept up with
great spirit. The young people seemed very happy.

——— ———

Thursday, 17th October.

Lady Florence Barnardiston and her sister Lady
Charlotte arrived. Fanny and all the party, except
Sarah Hervey, myself and Clement went to the
Bury ball and returned very late.

———

October 18th.

A succession of gaieties and a most delightful
party in the house from the beginning of this

1867. month.   Our dear Minnie and Sarah Napier, Kate
MacMurdo, Rose Kingsley (a very charming and
admirable girl), Sarah and Kate Hervey (for part of
the time), and latterly the William Napiers and
many others.   Nothing special to record, but it
would be a very great mistake to suppose that one
is not benefited most importantly and in various
ways by the society of admirable women.

A beautiful day.   I walked round the arboretum
with Sir Geo. Young and Emily Napier.   Several
of our guests went away—also Clement.   The
Arthur Herveys dined with us.   Their two daughters
went back with them.

Read the Introduction and 1st chapter of
Dugald Stewart's " Philosophy of the Active and
Moral Power of Man."   Wrote some notes on ossil
Palms of the Miocene period from Herr and
Unger.

Tuesday, 22nd.

Sir Edmund Head and the Austins went away.
I attended the Quarter Sessions at Bury.   Took a
very pleasant walk with Sarah and Cissy.   Rose
and Sarah sang very agreeably.

LETTER.

Barton,
October 23rd, 1867.

My Dear Henry,
        You are quite right about the Fern which
you sent me as Lastrea Orcopteris, and I shall be

very much obliged to you for some roots of it. I 1867.
shall be thankful also for Polypodium phegopteris,
which we have not got, and which seems to be
generally more rare in cultivation than Dryopteris;
the latter succeeds very well here. The small Fern
you sent in your letter is as you supposed,
Asplenium ruta muraria. We have Rhamnus
catharticus (the true Buckthorn) in cultivation here,
and also growing wild in a hedge on Loft's farm;
but its berries are not showy, nor its flowers
either.

We have now very mild, moist, soft, relaxing
weather, without much rain however; and the
autumn colouring of the trees and shrubs is
gorgeous. It will not last long, for the leaves are
falling fast. I do not think I ever enjoyed the
beauty of the autumn more than this year. In spite
of the very sharp cold that we had two or three
weeks ago, the Belladonna Lilies have flowered in
great beauty this season, in the border in front of
the conservatory, and the Cobæa scandens trained
against the outside of the conservatory, is still in
flower.

I have paid the purchase money £960, and am
now rightful owner of the beerhouse and land, late
Cox's. Scott hopes to manage so that the property
may give me a fair interest for my money; if so,
it will be well: but I have hitherto looked on it as
an expenditure more necessary than agreeable:
to keep off a nuisance, not to gain a benefit.

Our pleasant, very pleasant party is breaking up.
Dear Minnie and Sarah are just gone, with much

1867. sorrow on both sides at parting. The William
Napiers will stay, I hope, till the end of the
month, and Rose Kingsley with them. She is an
admirable girl.

Our dance on the 16th was as successful as
possible: everybody looked well, and seemed to
enjoy it. I danced Sir Roger de Coverley with
Lady Cullum, who was in great force.

Our girls, Sarah, Kate MacMurdo, Rose Kingsley,
Surah and Kate Hervey looked charming: and
Cissy and Susy looked very well too. Our total
number was 67. I did not go to the Bury ball,
but the young people reported very well of it.

Fanny, I am happy to say, seems very well after
all her exertions in the cause of the young. Since
then we have had a visit from Sir Edmund Head,
and from Mr. and Mrs. Charles Austin.

I do not know whether you ever met Mr. Austin:
he was a great friend of the Bullers, whom Cissy
will remember living at Troston; I often met them
there.   Aftewards   he made a great fortune by
Parliamentary practice, bought and settled on an
estate in East Suffolk, and has become a useful and
important man in the country ; he has married a
wife some 30 years younger than himself, very
handsome and very attractive, so that we are very
glad to have made her acquaintance. He is an
exceedingly clever man, and very well informed, and
a friend of Sir E. Head. I am very sorry indeed
to hear that you were so ill with your old enemy,
but I trust you will have no further visitation of it,
and will keep well through the winter. Much love

to Cissy and the dear children, and pray say to
Emmy and George that I am very much obliged to
both of them for the letters which they kindly wrote
to me a long time ago, and which I intended to
answer, but have not yet been able to manage it.

Ever your very affectionate Brother,

CHARLES J. F. BUNBURY.

I find I have libelled the weather, which is
beautiful, and almost as warm as summer. Cissy
will be more than ever indignant at the success of
Emperor Nap's bullying of the Italians. The Pope
appears to be safe till France is fairly at war with
Prussia: and then let him look out.

JOURNAL.

Thursday, 24th.

Read the article on Abyssinia in the new
number of the *Quarterly*.

It is by William Napier, and is very well done.
Very clear and sensible, without verbiage.

Mrs. Wilson and her eldest daughter and Miss
Milbank, Sarah and Kate Hervey and their two
young sisters spent the afternoon with us.

LETTER.

Barton,
October 24th, 67.

My Dear Katharine,

The *Shamrock* charade which you sent us
was very good. I found it out the same morning
that we received it.

1867.   We are looking forward to a quiet winter, and I plan great things in the way of study, of which, of course, I shall not accomplish one quarter, but never mind, it is good to propose to one's self a good wide scheme. In particular, I want to rub up my Greek, which has grown very rusty.

I see some promising articles in Murray's list of "forthcoming works" for this next winter. Motley's concluding volumes, Darwin's new book, and Lord Stanhope's "History of Queen Anne."

We have delightful weather now, and the colours of the leaves are quite charming. I do not think I ever enjoyed the autumnal beauty of the foliage more than this year. In spite of the sharp frost at the beginning of this month, too, there is still a considerable show of flowers out-of-doors : the Amaryllis belladonna has flowered beautifully this year, and is hardly yet over ; Cobæa scandens still in flower on the outside of the conservatory. The Ferns are doing well, and I expect some roots of Nephrodium Oreopteris from Abergwynant.

I am in perfect health, but some of our party are suffering from colds. With much love to Harry and your children,

I am ever,

Your very affectionate brother,

CHARLES J. F. BUNBURY.

Read the article on Abyssinia in the new number of the *Quarterly*, it is by William Napier, and very good.

JOURNAL.

Went into Bury to a meeting of the militia 1867. committee at the barracks.

————

Sunday, October 27th.

Read the article on the correspondence with Napoleon in the new number of the *Edinburgh*. Very interesting and instructive. A powerful vigorous exposure and dissection of his character based entirely on his own letters. It thoroughly bears out the opinion which I had always formed of the great conqueror, that he was as great as a man can be who has no moral greatness.

══════

LETTER.

Barton,
October 31st, 1867.

My Dear Edward,

I was very glad to receive your letter yesterday afternoon, and to hear of your safe arrival, though sorry that your tour should have been so much spoilt by bad weather. I was much entertained by your first letter, written from a place of which I had never heard the name; however I succeeded in finding this Klon Thal in the maps and tracing your wanderings so far. It must have been very pleasant to find out places entirely exempt from the usual deluge of English tourists,

1867. and I can well imagine the scenery being very interesting. Engelberg I perfectly remember.

Here we had extremely cold weather about the end of September and the first week or so of this month, which I suppose was about the same time that you met with so much snow in Switzerland. The autumnal colouring has been even more than usually beautiful this year, or at least I think I have enjoyed it even more than usual; but now the trees (oaks excepted) are in great part stripped.

Since our ball party broke up, I have settled again tolerably well to something of study. I want, in the course of this winter, to rub up my Greek, which has grown sadly rusty, so I have begun the "Odyssey," and have just got through the first book. I hope I may have perseverance enough to go through the whole, and then to try some of Plato.

I have not yet finished Burton, but having got into the reign of Mary Stuart, I find his narrative more flowing and interesting than in what went before; in treating of John Knox, he is even animated. The reigns of the Jameses, he treats in a very dry and meagre way; I do not think he has a particle of sympathy with anything romantic. In fact one might almost suppose that he had meant all the previous part of his history to be merely introductory to the reign of Mary.

In the new number of the *Quarterly*, the article on Abyssinia is by William Napier; that on the Talmud (I understand) by Max Müller; that on "The Retreat from Moscow," by Lord Stanhope.

I should like to know *who* wrote that in the *Edin-* 1867. *burgh* on the Napoleon correspondence ; I think it very good. I am looking forward to reading Darwin's new book, and Motley's concluding volumes, which I see announced.

We have had a housefull all this month, and very pleasant company ; I will not give you a list of names. But now all our guests are dispersed.

We are thinking of going to London about the 22nd of November, for a short time. The state of things in Italy is indeed most interesting and extraordinary. I hardly know what to expect, hardly even what to hope for. I certainly am very far from wishing for the maintenance of the Pope's temporal power, yet I am afraid that Garibaldi and his friends are very rash, and have chosen their time very ill.

Ever your very affectionate brother,
CHARLES J. F. BUNBURY.

JOURNAL.

November 1st.

Read chapter 39 of Burton's "Scotland," a short survey of the condition of the nation from the War of Independence to the Reformation—good. What relates to education is interesting, especially indications of the unusual attention paid to education in Scotland even before the Reformation. Burton's remarks on the University system as it existed in that country and on the Continent in the 15th and

1867. 16th centuries, are striking, even eloquent, and he points out with great force, the difference between the University system and that of colleges which in England grew to supersede the other.

* * *

November 2nd.

Read an article on "The Birds of Norfolk," in *Fraser's Magazine*—very good: written, if I am not mistaken, by Charles Kingsley.*

===

LETTER.

Barton,
November 3rd.

My Dear Henry,

I am very much obliged to you for the information you sent me about our father's commission and steps of promotion in the army. Some of them indeed, I had already ascertained, either from a slight outline, which he left of his military career, or from the Annual Register; but I am very glad to hear the complete list. It was in the year 1800, while residing with the Duke of York, as his A. D. C., that he took the resolution to improve himself and make himself fit for important commands, and with this view obtained leave to go and study at the military college. I always think it was a most remarkable proof of the energy and greatness of his character, that with no advantages

---

* Begun by Alfred Newton, but principally written by Charles Kingsley.

of education or example, after living in such an idle 1867.
and dissipated set as that at Oatlands, and
with every temptation to continue in a course
of idleness and dissipation, he should of his
own accord have formed such a resolution* and
adhered to it with such steadiness that in a very few
years after, he earned the esteem and confidence of
Sir James Craig and Sir John Moore. I am very
sorry I have not been able to find among his notes
anything relating to his studies or his companions
at the military college. Do you happen to remember anything that he may have said about it?

We are now quite alone; Edward left us yesterday morning; and I hope we shall be alone and
undisturbed for a fortnight to come, that Fanny
may have time to recover from her fatigues. We
had not been alone for a day since the 21st August.
We have plenty of agreeable occupation.

I trust you and dear Cissy and the chicks are
quite well. With much love,

    Believe me ever,
        Your very affectionate brother,
            CHARLES J. F. BUNBURY.

JOURNAL.

Sunday, November 3rd.

Read the article in the last number of the
Quarterly, on the French Retreat from Moscow.
Written I understand by Lord Stanhope—most
painfully interesting.

---

* It was from reading the thoughts of Marcus Antoninus.

November 4th.

Read part of chapter 40 of Burton's "Scotland,"
on the literature and language of the nation from
early times to the reign of Queen Mary:—curious
and interesting. I am particularly struck by
observing that this writer who is so far from
enthusiastic or credulous, believes in the antiquity
of Celtic poetry. The "Ossianic" poems—though of
course not in the exact forms in which they were
given by MacPherson.

———

November 5th.

Cecil went away. Read the remainder of
Burton's "Scotland," chapter 40. An interesting
notice of the architecture, ecclesiastical and
baronial of that country down to the time of the
Reformation. It appears that the earlier remains of
cathedrals and monastic buildings of Scotland,
those of earlier date than the War of Independence
are in the Norman and early English styles, closely
agreeing with those of corresponding periods of
England, but during the greater part of the 14th
and 15th centuries, the people incessantly engaged
in defending themselves against the English, had
not the means or leisure to build fine churches.
There are, it appears, few traces of anything
corresponding to the decorated in England. And
when the Scottish ecclesiastical architecture re-
appears towards the end of the 15th century, it has
a decidedly French type; it is not our perpendicular
but the French "flamboyant." The baronial
architecture Burton remarks, shows strong signs of

the unsettled state of the country, the change from castles to mansions took place much later in Scotland than in England.

———

November 7th.

Beautiful weather. We heard of the sad death of Mr. Weight.*

———

Monday, 11th November.

We went to Mildenhall, and I attended a meeting of the Literary Institute there.

———

November 12th, Mildenhall.

I went with Scott to West Row. Saw the new cottages building there.—Satisfactory.—Also the new farm buildings at Mr. Dawling's. We dined with poor old Mrs. Bucke who has nearly completed her 85th year. We returned home.

═══

LETTER.

Barton,
November 16th, 1867.

My Dear Edward,

Yesterday, Kate and I dined with the Wilson's at Stowlangtoft, and met Lord and Lady Bristol and Lady Mary, the Charles Austins, Lady Cullum, Charles Tyrell and his brother, the Major, &c.:—a pleasant party.

I am much obliged to you for your information

* The man who collected his London rents.

1867. about the Greek Lexicon, and will certainly order
the one you recommend : but in the meantime
I do not use Donnegan, but Dammii Lexicon
Homericum, the Glasgow edition of 1825, in two
volumes, 8vo.,—a present by the way, from you,
in 1837. I find it very useful. It so happens
too that the copy of the Odyssey which I am
now reading was a gift from you : it is a
corpulent little volume, published at Liepsic
in 1828. I make use sometimes too of old Clarke's
"Homer." I cannot say that I find the first two
books of the Odyssey very interesting, but no doubt
it will improve as I go on.

We still expect to get to London about the end
of this month. Fanny sends her love.

Ever your affectionate brother,

CHARLES J. F. BUNBURY.

JOURNAL.

Sunday, November 17th.

Read chapter 45 of Burton, coming down to the
murder of Darnley. Burton improves very much
as I go on. This 4th vol. is much better written
and in a more interesting manner than the others,
and the latter part of this volume better than the
earlier. His manner is as different as possible
from Froude's, but the murder of Rizzio and that
of Darnley are very well told, in a grave, sedate,
clear, simple manner, and with a circumstantial
precision which are sufficiently impressive. It is
something like the style in which a judge might

expound an interesting and intricate case in a 1867.
criminal trial to a jury.

Monday, November 18th.

A startling telegraphic message in the newspapers
yesterday, that the Island of Tortola, one of the
Virgin Islands (in the West Indies) had been
"submerged" and ten thousand lives lost. It is
hardly possible it can be true, as the Island consists
(it is said) of a mass of rocky hills, rising in one
part the height of 1600 feet, and a sinking down or
submergence of such a mass would be a geological
miracle to which there has been nothing parallel
within historical times. Besides, the whole population of the Island it seems was scarcely 6,500.
But I am very anxious to know if there is any
foundation at all for the report.*

Friday, 22nd.

Read chapter 47 of Burton's "Scotland," concluding the 4th vol., the last yet published. It ends
with the Abdication of Mary, July, 1567. In this
chapter he gives a very full and particular analysis
of the famous casket letters. He evidently believes
in their genuiness, though he does not absolutely
affirm it but admits the difficulties respecting them.
It would seem that the external evidence is imperfect and unsatisfactory, but the internal evidence

* NOTE.—*November 24th.* This report concerning Tortola has, as I expected,
been contradicted by the official statement of the governor; but the *Town*
seems to have been almost destroyed, and the unfortunate inhabitants totally
ruined. The hurricane at the neighbouring island of St. Thomas appears to
have been one of the most dreadfully destructive on record.—(C. J. F B.)

1867. very strong,   Burton makes much the same remark
on them which Kingsley made to me in conversation,—that if they were not genuine it was
difficult to conjecture *who* was capable of composing
them: if they were not the real spontaneous out-
pourings of a woman's heart and mind, the forger of
them   must   have   been   almost   a   Shakespeare.
The only possible person who could be named as
capable of forging them would be Buchanan, and
Burton thinks them totally unlike his manner.
Altogether this fourth volume of Burton is very
interesting, and to my thinking a very good history.
The   story   of   Mary Stuart   and   her   lovers   and
husbands is always interesting : one may read many
versions of it without risk of being tired, and here it
is very ably and judiciously treated.   I might say
judicially as well as *judiciously*, for there is much of
the temperate, cool, guarded spirit of a judge in
Burton's manner of handling the topics.

It would seem as if he had intended the three
previous   volumes   to   contain   merely   a   sketch
introductory to the history of this reign.   There is a
great contrast between his dry and meagre treatment of the reigns of the James's and the interesting
fullness and copiousness of this volume.

Monday, 25th.

A visit from Arthur Hervey, Sarah and Kate.

Wednesday, 27th.

A beautiful day.   Preparing for our journey
to London.   Yesterday I wrote the conclusion of

my Memoir of my father's life.   I began it on the   1867.
7th August, 1866.   But this is only the first rough
draught ; if I live I purpose to devote some months
to a very careful revision and correction of the
whole ; and after all I apprehend it will only be
tolerable.   I do not feel that I have any talent for
biography.

A beautiful morning.   Up to London.   Found
our house very comfortable.

Friday, 29th.

I went to see poor George Napier ; found him
recovering from his severe illness — then to the
Athenæum and to see Edward.

The Charles Lyells and Katharine and William
Napier came to afternoon tea with us, and afterwards the Arthur Herveys and Edward.

Monday, 2nd December.

Snow.   Bitterly cold weather.   I went to Harley
street and had luncheon with the Lyells.   Dear
Kate MacMurdo came back from Wimbledon.

Edward dined with us.

Tuesday, 3rd.

Miserably cold.   I began to read the third volume
(newly published) of Motley's "History of the
United Netherlands."   The story of the wonderful
surprise of Breda by the Dutch, February, 1590, is

1867. exceedingly interesting. We went out to Norwood
and visited Susan and Joanna in their new house.

_____

Thursday, 5th.

Weather milder. Attended the Council of the
Linnean Society. Had some talk with Bentham
and Hooker. Montague MacMurdo arrived unexpectedly while we were at dinner, and spent part of
the evening with us.

_____

Friday 6th.

We received news of the death of poor Wm.
Blake, our Vicar. The MacMurdos came to
luncheon with us, and had dinner also at our house
and then set off for France.

We dined with the Henry Lyells, and coming
away about eleven o'clock, immediately perceived a
glare in the sky towards the south-east, which
showed that there was a great fire. The reflected
light on the clouds in that quarter was strangely
vivid. After a time we learnt that it was the opera
house which was on fire; so driving down to the
lower end of the Regent Quadrant, whence we saw
directly down Waterloo Place, we stayed a good
while in our carriage looking on at the conflagration.
It was the first *great* fire of which I ever had a near
view, and it was an astonishing sight. The
enormous volumes of smoke intensely coloured by
fire bursting up incessantly with an impetuosity as
if shot up by a volcano, and rolling away mass over
mass towards the south; the fresh glare sometimes
of yellow, sometimes of deep red blazing up from

time to time, the black appearance of the houses 1867.
between us and the fire, made a sublime and terrible
sight.

———— ———

Dreadful weather.

Edward, who saw the fire from the balcony
of the Athenæum, says that even there, when the
fire was at its height, the heat was almost unbear-
able. Little Harry came to stay with us, and
Katharine with Arthur and Rosamond came to
luncheon. Fanny and Kate went out with them.
I wrote to Patrick Blake.*

—— ————

I have since the 3rd December read seven
more chapters of Motley's "Netherlands." There
is in many places a disagreeably controversial tone, a
soreness and irritability apparent in his manner of
writing, a laboured effort to exalt the cause of
republicanism, which detracts much from my
pleasure in reading him, but his narrative is as in
the former volumes, excellent, and the history is in
itself, extremely interesting. The hero of it undoubtedly is Maurice of Nassau, whose military
genius and ability improved by assiduous study,
were something really wonderful. Alexander
Farnese and his campaigns against Henry IV. of
France, introduce a romantic and chivalrous
element into the story. It was most fortunate for
the Netherlands, and indeed for Europe, and for
mankind in general, that Philip of Spain set his

* Admiral Blake, brother of the Vicar of Barton, who had lately died.

1867. heart on conquering France, and therefore (after
the murder of Henry III. in r589) turned his energies
in that direction to the neglect of the Netherlands.

————

We dined with Charles Lyell.   Met Mr. Twisleton, Murchison, the Benthams, Lady Bell, Mr.
Franks.   Mr. Twisleton talking of Sir James
Mackintosh (to whom he thinks Sir Henry Bulwer
in his his "Historical Characters" has scarcely
done justice) said Parr's famous sarcasm on Mackintosh is not adequately given in the *Quarterly
Review* article on that work.   As he (Mr. T.)
had heard it, Mackintosh and Parr were in company
just at the time when the former was accused of
deserting his political friends by accepting the
appointment to Bombay.   The fate and character
of Quigley, the Irish rebel, was discussed, and
Mackintosh ended by saying — "After all he was
"as bad as he could be ; he was an Irishman, a
"priest and a traitor."   "No, Jemmy," said Parr,
"I do not agree with you—you say he was as bad
"as he could be, I say he might have been worse—
"you say he was an Irishman, he might have been
"a Scotchman—you say he was a priest ; he might
"have been a lawyer—you say he was a traitor ; he
"might have been an apostate."
Mr. Twisleton told this story with great effect.
Mr. Twisleton said that if Mackintosh had been
a more selfish man, and more disposed to self-
assertion, he might have gained more fame as well
as power.   He said that Brougham was the person

aimed at by Sydney Smith as a contrast to Mackin- 1867.
tosh, in the passage beginning "if he had been
arrogant and grasping, if he had been faithless and
false." (See "Life of Mackintosh." vol. 2. p. 503).

We had talk about Junius's letters. Mr. Franks
said he could affirm that Junius and Sir Philip
Francis used the same seal. It seems he has
made a particular study of these subjects—seals,
engraved stones and the like—and Mr. Parkes (who
began "The Life of Sir P. Francis," which has been
completed by Herman Merivale) showed him all the
seals which have been preserved of the original
letters of Junius.

He satisfied himself by minute examination,
although pains had been taken to render the
impressions indistinct, that some of them had been
made with the official seal of Lord Barrington,
which was one used by Francis, and that others
would be identified with another seal (an antique
head, I think) which was also used by Francis.

Mr. Twisleton said that Lord Holland used to
deny that Francis could have been the author
of Junius on this ground. That he was so vain a
man that it would have been impossible for him
to have avoided pluming himself on such a work,
if he could have claimed it.

Mary had in the evening, Miss Shirreff, Julia
Moore, Mr. Zincke, Mrs. Young and her daughters,
the Louis Mallets and others.

Friday, 13th December.

We went to Salviati to order a chandelier.

1867.   Saturday, 14th.

News of the extraordinary Fenian crime at
Clerkenwell. Visit from Augusta Freeman and
from the Wm. Napiers. Went on reading Motley.
I have now come to the 32nd chapter, the 12th of
this volume, the arrival of the Archduke Cardinal
Albert, as governor in the "obedient" provinces. As
I said before, I am often bored by the authors
obtrusive republicanism which he is so continually
and so unnecessarily thrusting in our faces. But
the narrative is full of interest and very well done,
especially in the military incidents. The satirical
description of the absurd allegorical pageantry got
up to welcome the Archduke Ernest into Brussels
(chap. 30) is very amusing. The book is, perhaps,
the more interesting because Motley identifies himself so much with the Hollanders, and looks at the
history so much from their point of view, that he is
hardly capable of doing justice to Henry IV. or
Elizabeth, still less to the Spaniards. This par-
tisanship makes him write with more spirit and fire.

———— —

Monday, 16th December.

We had a visit and a long talk with the Rev.
Mr. Percy Smith, formerly Kingsley's curate.
Sir Edmund Head, George Napier, and the
Henry Lyells dined with us. Henry arrived in the
evening.

————

Tuesday, 17th December.

Went to the Zoological Gardens ; saw the young

Walrus : it was lying motionless on the gravel walk, 1867. looking very lethargic.

At present its appearance is far from striking or interesting, looks very fat, clumsy or shapeless, but not grotesque and ridiculous, as in the old prints ; its tusks not yet visible; its huge whiskers, the most peculiar thing about it ; the skin covered with short, close smooth hair of a dirty brown colour.

The common Seals in the next enclosure to it, much more interesting animals, very lively and active, with fine black eyes and intelligent dog-like countenances. Their motions on dry land very odd, but surprisingly fast, though apparently awkward.

In another part of the gardens, a fine Ant-eater, certainly one of the most extraordinary of all quadrupeds ; the formation of its head and snout so wonderfully strange, as to appear unnatural even in the living animal ; its feet hardly less peculiar, for it walks resting (seemingly) on the back of its paws, its toes and huge claws turned completely inwards, so that its feet look like stumps; its enormous bushy tail like a banner, and the peculiar arrangement of colour complete the oddity of its appearance.

In the Monkey house are two Chimpanzees, which walk about with their arms round each other in most loving fashion, with a sort of comic pathetic expression in their faces.

The William Napiers, George Napier, and Edward dined with ns.

———

1867.   Tuesday, 18th December.

Packing up. Visits from Susan and Joanna,
Edward and Henry.

––––––––

Thursday, 19th December.

A sharp white frost, bitterly cold. We (with Kate
MacMurdo) travelled by train down to Brandon,
and from thence in a fly to Weeting,—Mr. Angerstein's. Admiral and Lady Harriet Baillie Hamilton
with their son and daughter, arrived at the same
time.

I sat at dinner between Mr. Angerstein and Cissy
Ellice.

––––––––

Friday, 20th December.

A bright day and hard frost.

I took a walk by myself and saw the remains
of the old castle. In the afternoon took a walk with
Mrs. Angerstein ; had pleasant talk with her at
dinner, and with Admiral Baillie Hamilton in the
evening.

––––––––

Saturday, 21st December,

A wretched day ; thaw and heavy rain. We left
Weeting and returned by Brandon and Thetford to
our dear home. All safe and well at home. Thank
God.

I read in Motley the very spirited narrative of the
remarkable action at Turnhout, (January 24th,
1597, in which Maurice of Nassau with only 800
cavalry and no infantry, utterly cut up and de-

molished four of the best infantry regiments in the 1867. Spanish service. It is exceedingly well and clearly told. Motley, however, seems to acknowledge that it was a rash exploit, that all the chances were against Maurice.

Monday, 23rd December.

A visit from Patrick Blake, to talk about his brother. He was much overcome.*

Thursday, 26th December.

Wrote to Mr. Babington (Cockfield) sending him money for the charities of his parish. We went to Ickworth and had luncheon with the Arthur Herveys.

Friday, 27th December.

Read in Motley the account of the death of Philip II.

It is an impressive and almost awful instance of self-deception, that that man should on his death-bed have said to his confessor (and seemingly with sincerity) that he could not recollect having ever in all his life, wilfully wronged or injured any man. It shows how thoroughly the conscience may be seared and perverted by bigotry. Yet I can hardly believe that a *good-hearted* man could under any system of belief have done what Philip did. I suppose that all the personal vices of Nero, and of all the worst Roman Emperors', did not cause one

---

* Admiral Blake, who had just lost his brother, Rev. William, Vicar of Barton

1867. half the misery to mankind which the sincere bigotry of Philip did. No, I am not sure that I ought to lay it all to his bigotry ; I believe that he was fully as much a despot as a bigot ; that he persecuted the Protestants quite as much because they opposed his will, as because they opposed what he thought the true religion.

———— ——

Saturday, 28th December.

Went to Risby and saw the Abrahams.

———— ——

LETTER.

Barton,
December 30th, 1867.

My Dear Katharine,

With all my heart I wish you a happy, very happy new year, and many of them, to yourself and all your loved ones. May 1868 bring you every blessing, and may its close find us all with as many reasons for thankfulness as its beginning. I cannot help being sorry for the departure of the old year. I look back on its course with a very warm sense of gratitude for many comforts and blessings, and much happiness enjoyed, with very few drawbacks ; I cannot venture to hope that the years to come will all be as bright, but I trust that whatever happens will be right and for our good. When we look beyond the circle of our own concerns, and our immediate surroundings, the prospect is not bright ; indeed, I think the political prospect in all directions at home and abroad, is about as dark and 1867. unpleasant as I almost ever remember ; we can only trust that Providence will bring good out of evil.

Fanny has been even more than usually busy since we returned home. The weather has, for the most part, been either so cold or so dark, or both, that I have been able to make hardly any use of my museum ; not even to make much progress in arranging those beautiful specimens of Ferns which you so kindly gave me. I have been busy reading old letters of my father which Cissy has lent me, and making extracts, to be worked into their places in my Memoir.

I go on reading Motley, which is very interesting, and in the evenings I read Shakespeare aloud to Fanny and Katie. On two evenings also I read to them Washington Irving's delightful chapters on "Christmas," in the Sketch Book. How charming his style is.

(31st). We have had a "green Yule," as yet, but it is bitterly cold, and looks quite as if snow were coming. Yesterday I tried to look for mosses, but got my face nipped by the bitter wind, so that to-day I have had a threatening of tooth-ache, and therefore have stayed at home.

Our ferns (in the houses) are looking well ; and the conservatory is gay with Justicia speciosa, Sericographis, Jasminum nudiflorum, Linum trigynum, three kinds of Begonia (one particularly pretty), Celosia, Cyclamen, Primula Sinensis, and other pretty things. We are expecting some gaieties in the middle of January.

1867.   With much love and every good wish to all your family.

Ever your very affectionate brother.

CHARLES J. F. BUNBURY.

JOURNAL.

Tuesday, 31st December.

Lady Cullum came to an early dinner, and spent the evening with us.

The servants' dance and merry-making.

So ends 1867. Thanks to God for all the blessings enjoyed in it.

1868.

Wednesday, 1st January.

1868.   Lady Cullum, who was staying with us, went away after breakfast. Had a pleasant visit from Arthur Hervey. Poor Scott's baby died. I began a careful revision and correction of the Memoir of my father.

Thursday, 2nd January.

Hard frost and snow. I finished reading vol. iii. of Motley's "History of the United Netherlands." The chapter on the Maritime Expeditions and Discoveries of the Dutch is exceedingly interesting, especially the narrative of the wonderful enterprise

of Barendz and Heemskerk, when they wintered on 1868.
the coast of Nova Zembla. It seems almost
inconceivable that any human beings could have
survived such hardships and sufferings.

The labourers' supper party. Fanny and Kate
went to a dance at Ampton.

Frost and a little snow. Fanny and Kate went
to a dance at the Wilsons, and stayed late.

Much business in the morning. Sarah and Kate
Hervey came to luncheon, and spent part of the
afternoon with us. Read chapter 48, the first of the
4th volume, of Motley's "United Netherlands:"
the furious battle on the sand hills (dunes) at
Nieuport: related very fully and with remarkable
clearness and spirit. The turns of fortune in this
battle were extraordinary. The Dutch were on the
very point of losing it, and they were in such a
situation that a defeat must have been the destruction of that army, and very probably the ruin of
the republic. The expedition was, in its very
conception, as Motley shews, an enormous error:
but this was not Maurice's fault, it was forced upon
him by the States-General, and he did everything
that could be done to redeem the blunder. The
armies of that time were very small compared with
those of modern days. This battle of Nieuport was
one of the most important of the whole long war:

1868. yet it seems that neither side brought into the line of battle much more than twelve thousand men. Patrick Blake dined with us.

———

Monday, 6th.

Snowing most part of the day, but scarcely freezing. Read the Odyssey to end of book 3. In this book are many pleasing and interesting touches of manner and character.

———

Wednesday, 8th.

Mr. Percy Smith arrived.

———

Thursday, 9th.

Poor Scott's wife died, to our great sorrow. Mr. Smith went round the parish with Dudley Clarke, and saw the Vicarage, Church, &c.

———

Friday, 10th.

Mr. Smith visited poor Scott, and also John Phillips. We agreed finally about the living; he left us in the afternoon, leaving us a very pleasant impression of him.

———

Saturday, 11th.

I filled up, signed, and sent off to Mr. Percy Smith, the presentation to the living of Barton. Finished chapter 40 of Motley; continuation of the Siege of Ostend, including a very spirited assault, very well described: the naval victories of the Dutch, their progress in the East Indies, the establishment of an East India Company, March 1602.

We received the news of poor Richard Napier's death, which happened yesterday morning at half-past four, after a very short illness. He was in such a melancholy condition, widowed, blind and lonely, that death was a happy and merciful release for him. Longer life would have been only a prolongation of suffering. But my thoughts recur to the times when his mind and body were in vigour, to the many long walks and delightful conversations we used to have together. He was always, from my childhood, most kind to me and seemed to like my company, and I loved him much, as well I might. He was, I think, one of the best men in the world: a warmer or more loving heart, a more generous and noble nature never existed, but owing to the extreme tenderness and sensitiveness of his disposition, he seemed to be on the whole, not as happy as so good a man should have been. Perhaps too, he had a constitutional tendency to melancholy. When in good spirits he was a most delightful companion. His knowledge was very extensive and various, his memory excellent, and particularly well-stored with the best poetry, and he had a greater flow of perfectly unaffected and unstudied eloquence than any other man I have ever known. He had also a great stock of good stories, and told them with much humour.

Dear Minnie and Sarah arrived and Helen Ellice with them.

Wednesday, 15th January.

A beautiful mild day. I was a good deal out-of-doors about the grounds. Mrs. and Miss Berners arrived and John Hervey.

The *Pall Mall Gazette* reports that the famous Dragon Tree of Orotava was shattered to pieces by a gale of wind last autumn. I am glad to have seen it. It was one of the most remarkable trees in the world.

––––

Thursday, 16th January.

Fanny with Mrs. Berners and the young people went to Lady Bristol's ball.

I read chapter 43 of Motley, concluding the great Siege of Ostend. It was certainly a wonderful siege which lasted 3 years and 77 days, and came to an end at last only when the town and its defences had been so utterly destroyed that nothing remained to be defended. Like the Siege of Sebastopol, however, it was not a regular siege—not a siege in the strictest sense: the town was never invested, its supplies never cut off. The desolate scene it exhibited at the time of its surrender is strikingly and powerfully described by Motley.

––––

17th January. Friday.

A talk with poor Scott, who bears his great misfortune bravely. Fanny, Mrs. Berners and the young people went to the Bury ball.

––––

Saturday, 18th January. 1868.

Furious wind and rain. Mrs. and Miss Berners went away. Lady Rowley and two daughters came to luncheon with us.

LETTER.

Barton.
January 18th, 68.

My Dear Katharine,

Poor dear Richard Napier's death ought hardly to be an occasion of mourning; it is to him such a happy and welcome release from a world in which he had nothing left to live for, that I feel one ought rather to rejoice at his liberation. For the last year he had been, in effect. dead to all but sorrow. I think often and with pleasure which will never fail, of the former bright days when I had so many delightful conversations with him. When he was in good spirits he was one of the most delightful companions I ever knew: such a wonderful flow of unaffected eloquence, such power of language, such extensive and various knowledge, and with all this, so much humour and such a fund of good stories. He was one of the kindest, warmest-hearted men that ever lived, and I do believe one of the best: though over sensitive, and therefore not quite so happy, perhaps, as so good a man should have been. He was, indeed, a man not to be forgotten by those who have ever known him.

Poor Scott's loss of his wife is indeed a terrible

1868. misfortune. It would be difficult to imagine a heavier blow falling on any man: but he bears it like a brave man, and an excellently good man as he is. I hope his eldest daughter will be a comfort and support to him.

I have been much interested in hearing of dear Leonard's examination; I have little doubt that he will be successful, because I believe him to be clever, and I know that he is exceedingly diligent, earnest and thoughtful, — a conscientious, hard worker.

Fanny is taking a good rest to-day (Sunday the 19th) being very much tired, as well she may be, by chaperoning young ladies to two balls in succession. However, I trust she is not really the worse.

We have dear Minnie and Sarah and Miss Helen Ellice with us now; and in the latter part of the week we expect a learned party:—Joseph Hooker, Alfred Newton (Professor of Zoology at Cambridge) and Mr. Churchill Babington.

With much love to all your family party, believe me ever,

Your very affectionate Brother,
CHARLES J. F. BUNBURY.

JOURNAL.

Monday, 20th Jaauary.

Helen Ellice went away. Read the Odyssey, v. 138-218 of book 4. There is something amusingly quaint in the simplicity with which Peisistratus says 1868. that he does not much like to weep and lament at supper time.

The Arthur Herveys, Lady Cullum, Mr. Bevan and Mr. Lott dined with us.

Tuesday, 21st January.

Dear Minnie and Sarah went away.

Wednesday, 22nd January.

Froude remarks very justly (Short Studies) there is something wild and mysterious in the story of the Trojan Horse, as told in the Odyssey, and especially in Helen's share in it.

Friday, 24th January.

Hard frost in the morning. Joseph Hooker and Alfred Newton arrived. I showed them the arboretum and the ferns: afterwards Mr. Churchill Babington arrived. Arthur Hervey and dear Sarah and Kate dined with us.

Saturday, 25th January.

We took a walk with Hooker and Newton and Mr. Babington to the stables, arboretum, ferns, &c. Mr. Babington went away after luncheon. Dr. Hooker very agreeable and interesting.—Alfred Newton also very pleasant.

Sunday, January 26th.

I took a walk with Joseph Hooker and Alfred

1868. Newton.   We showed them the autographs, and
had much pleasant talk.

———

Dear Katie MacMurdo left us to return to her
family, a sorrowful parting.   Hooker and Newton
also went away.

Joseph Hooker and Alfred Newton (the Cambridge
Professor of Zoology) had stayed with us from the
24th; both very pleasant.

Hooker is an admirable man, and a very interesting one ; I have known him now many years, and
always admired him, and I think he improves still
further upon further acquaintance.   My acquaintance with him began April 3rd, 1846, at Mr.
Horner's house, in Bedford Place.   He is not only
one of the greatest botanists now living, but has a
great variety of information and of pursuits ; a
generally well cultivated and remarkably active
intellect.   An eager, impetuous nature, somewhat
excitable, I should think, though capable of a vast
amount of work.   He is, in natural science, a keen
Darwinian, and in general a warm advocate of what
are called "liberal" and "progressive" doctrines,
though not violent or extravagant ; not a *subversionist*
like Huxley.   Eager to welcome new discoveries,
and to follow up new thoughts and suggestions, he
is at the same time not at all deficient in veneration,
and is able and willing to do full justice to the
learned and the good of former times.   Hooker said
that what he thought especially remarkable and
admirable in Lyell as a man of science, was his

candour and fairness ; his anxious care to under- 1868.
stand thoroughly the opinions and arguments of his
opponents, and to state them fully and fairly. In
this I entirely agree with him.

He thinks that Lyell's complete conversion,
and open avowal of his conversion to the Darwinian theory, at his time of life, and with his
established celebrity, and after he had elaborately
argued against the same theory in many editions of
his great work, is a phenomenon almost unexampled
in science.

Hooker wishes for the separation of the natural
history collections, now in the British Museum,
from the rest, but would not send them to South
Kensington. He is for establishing them near the
southern part of the Regent's Park. Bloomsbury,
he said, is a bad situation, owing to the want of
light. He is much against the system of trustees,
and would have every institution placed under a
single responsible head ; in which I heartily agree
with him.

He admired many of our trees, especially the
Cephalonian Fir, the Cryptomeria (the one in the
arboretum), Pinus excelsa and Magnolia acuminata.
Those at Hardwick he admired still more, and said
that Lady Cullum's Conifers, especially the
Deodaras and Araucarias, are far superior to the
boasted ones at Elvaston.

He observes that the Wellingtonia makes repeated
growths in the year, whence it is more difficult
than any other Conifer to distinguish the shoot of
one year from that of the past; therefore he suspects

1868. that more than one ring of growth may be formed in one year, and thus that the estimates of its enormous age, may be fallacious. For other notes of Hooker's remarks on botany, see my natural history note book.

Joseph Hooker told me that Charles Darwin offered to give his new book to his brother, if he would promise to read it. " No, I thank you," said Erasmus Darwin, "I would rather buy it than read it."

Newton is a very good naturalist, and a well informed and very pleasant man. Birds are his especial study, and on them he is, I believe, a great authority. He has visited Iceland in pursuit of them, and it was amusing to hear him and Hooker comparing their Arctic and Antarctic experiences.

We drank tea with the Arthur Herveys. Herbert arrived.

January 28th.

We went to hear Arthur Hervey's second lecture on Napoleon, at the Bury Athenæum.

29th January. Wednesday.

Much shocked by hearing of the sudden death of Sir Edmund Head. He is a very great loss. It is only within the last few years that I have come to know him intimately, but in these last three years (especially) we have seen a great deal of him, and liked him exceedingly. He was most friendly and cordial to us, and seemed to enjoy his visits to Barton. I have scarcely known a man of more extensive and varied knowledge, or of a more powerful grasp of mind, and with this he had a remarkably refined taste and feeling for poetry. On a slight acquaintance his manner struck one as dry. I was quite surprised when I came to know him better, to find him full of humour and fun,— often as merry and light-hearted as a boy. He was a thoroughly genial man, and a most kind-hearted one.

Engaged a long time at Bury at a meeting concerning a proposed association for nurses.

LETTER.

Barton,
January 29th, 1868.

My Dear Edward,

I am very much shocked and grieved and Fanny still more so, at hearing of the death of Sir Edmund Head, but I thank you for sending us the earliest information of it. He is a very great loss. We have seen so much of him in the last few years, and grown so intimate with him, that we feel it particularly; the more I saw of him the more I liked him. It makes me very sad to think we shall have no more of his conversation, always so pleasant and so instructive. He was certainly an uncommonly agreeable man, and in every way a valuable one, and I should think his loss will be much felt in the public service. But poor Lady Head! what a dreadful irreparable blow to her. They were actually engaged to come here on the 12th of next month, and Fanny heard from Lady

1868. Head only the day before yesterday.   She (Fanny)
is quite shaken by this intelligence.   I had been for
some little time past, intending to write to you on
other subjects, but now I will wait a few days
longer.

Ever your very affectionate brother,
CHARLES J. F. BUNBURY.

JOURNAL.

31st January.

Fine weather.   A high west wind.   We went
with Scott through the Vicarage Grove to arrange
paths to be made and bushes to be cut, also to fix
a site of a new gate.   I received a packet of seeds
from Joseph Hooker.

Read chapter 47 of Motley:—a very remarkable
sea flight in the bay of Gibraltar, between a Dutch
fleet under Heemskerk (the same who wintered in
Nova Zembla) and a Spanish fleet under D'Avila,
both Admirals killed; the Spanish fleet entirely
destroyed by a very inferior force.

Saturday, 1st February.

I received the warrant appointing me Sheriff.
Drove out with Fanny; we visited Mrs. Lockhart at
Horringer.   An indifferent account of Sir James
Simpson; then to Hardwicke, and saw Lady
Cullum—very pleasant as she always is.   Called on
Sir John Walsham, who was out.   Went on with
my Memoir.

Read chapter 48 of Motley—a striking picture of the miserable state of Spain under the nominal reign of Philip III. and the real government of the Duc de Lerma. The effects of the second Philip's system in Church and State were already making themselves felt to a great extent in the first ten years of the 17th century. We see in the picture drawn by Motley most of the features that are so vigorously worked up by Macaulay in treating of the time of Charles II. of Spain.

Monday, 3rd February.

Mr. Sparkes and Mr. Phillips came to me, and I took the oaths as Sheriff, appointing Mr. Sparkes my Under-sheriff. Went on with my study of Brazilian plants. Wrote to Charles Lyell.

Tuesday, 4th February.

My 59th birthday. Thanks be to God.
I received very kind and pleasant letters from Edward, Mary, Katharine, Cissy and Henry.

LETTERS.

Barton,
February, 4th, 1868.

My Dear Edward,

Very many thanks for your agreeable letter, as well as for your kind wishes. I am glad to

A A

1868. give my vote to Mr. Bruce. I have already given
one proxy on behalf of Mr John Bateson, who was
recommended to me by Lord Bristol, and who is
a cousin of Fanny's, as well as a large share-
holder ; but I had still one to spare.

I most heartily agree with you in all you say
about Sir Edmund Head and Richard Napier.

I forget whether you told that the article in the
Edinburgh on Sir Philip Francis, was by Mr. Reeve.
It entertained me a good deal. As to the *Junius*
question, the arguments appear so strangely
balanced, that my judicial faculties are not sufficient
to decide between them. I am rather inclined to
suggest, as a probable hypothesis, that the "author
of Junius" was *the devil. Aut Franciscus aut
Diabolus.* Who else was there who could have done
it ? I see in Roger's "Recollections" that Grattan
thought that Burke was Junius ; but surely Burke
was much too good a man.

I was much interested by the articles in the same
number of the *Edinburgh* on "Philip II." and "Don
Carlos," and on "De Fezenzac's Recollections of
the Grand Army."

I should be very much obliged to you if you could
give me any information about our father's *pension*—
what was its amount, and under what circumstances
he gave it up ? It was given to him, I presume, on
his retiring from office. I wish I could recollect
some of the things he used to tell about the dis-
turbed state of the country when he first settled at
Mildenhall. I remember that some of the gentlemen in this neighbourhood were asking for soldiers

to defend their houses, but I do not know whether 1868.
any soldiers were quartered at Mildenhall.   Do you
remember hearing ?

   I trust that the choice I have made of a clergyman
will turn out a benefit to the parish, and everything
I have heard leads me to believe so.   I have
tried earnestly to make an honest and wise choice
according to the best information I could get, and I
am glad you think I have done well.   Mr. Percy
Smith is to be instituted by the Bishop on Thursday next, and is then coming to us for a few days
to be *inducted* and to read himself in.   But unluckily
Mrs. Smith is not able to come with him, I am
rather anxious tomake her acquaintance.

Ever your very affectionate brother,

CHARLES J. F. BUNBURY.

Barton,

February 5th, 1868.

My Dear Katharine,

         Very many thanks for your kind letter and
your good wishes, as well as for the beautiful
specimens of a very rare little fern, so neatly
mounted and labelled, which are a most welcome
acquisition to my collection on their own account,
beside their great value to me as tokens of your
kindness.

   Yes, indeed, the death of Sir Edmund Head is a
loss which we shall long feel.   It is not often that
one meets with such powers of mind, such a
vigorous understanding, and such extensive know-

1868. ledge, combined with such a kindly and genial nature, and such agreeableness in society. I quite agree with you, that people past the middle of life should on no account submit to an operation unless it be *absolutely* necessary.

Have you seen Dr. Hooker since he was here? He was exceedingly agreeable, and I was particularly glad to be able to show him the arboretum, &c. even though it was at such an unfavourable time of year. He seemed to admire some of our trees, but he admired Hardwick still more. Alfred Newton was also very pleasant.

We have been fortunate hitherto in not having any important trees damaged by the extraordinary tempests which have prevailed so much lately.

We are cutting walks through the vicarage grove, to give more convenient access to *my* tree, and Fanny's and Sarah's.

Do you see in the papers, the account of the great landslip at Naples? It is a terrible thing for the poor people, but rather curious.

I have just finished reading Sir Henry Bulwer's "Memoir of Talleyrand," the first volume of his "Historical Characters." Very agreeably written.

Much love to your husband and children.

Ever your very affectionate brother,
CHARLES J. F. BUNBURY.

JOURNAL.

We dined at Mr. Borton's: met the Abrahams, Lady Cullum, Sir Edward and Lady Gage, the Henry Blakes and Major Tyrell.

Very fine. Mr. Percy Smith arrived. I took a walk with him. We had a dinner party; Abrahams, Suttons and Bains. I read an admirable sermon in Kingsley's new volume.

Mr. Percy Smith was inducted into the Vicarage at Barton. Patrick Blake and Lady Cullum dined with us.

Very fine, but cold. Took a walk with Fanny. We went to afternoon church. Mr. Percy Smith read the service very well, and preached. The church well filled

---

Read the Odyssey, book 4, ver. 537-619. Telemachus very wisely declines Menelaus' proposed gift of horses, because Ithaca is not suited to them, having no plain ground fit for riding, nor pasturage to feed horses. He contrasts it in these

1868. respects with Lacedæmon—the meadows on the
banks of the Eurotas, where he says there is much
Lotus and Cypeiron as well as corn. Lotus here,
must evidently mean a different thing from the
Lotus of Egpyt, and from that of the Lotophagi:
must mean something that contributes to form
meadow herbage and horses' food: very probably
some kind of clover or trefoil. The *Kupeiron* in this
passage may very probably be the Cyperus longus,
which according to Sibthorpe (in Walpole's Turkey)
is still called *Kupeiron* in Greece.

February 11th.

A beautiful day. I was much out-of-doors enjoying the sunshine and the early flowers.

LETTER.

Barton,
February 13th, '68.

My Dear Katharine,
        Many thanks for your letter, which I
answer immediately, as you are going abroad so
soon.

I hope dear Frank will find himself comfortable
at Berlin, and that his stay there will answer all
your expectations; I have no doubt he will become
a good German scholar, and will see more variety
of men and manners than he would in London, and
pick up much knowledge, I am much in favour of
young men travelling, and seeing various countries:

" Homekeeping youth have ever homely wits."

I wish you joy most heartily of Leonard's success 1868. in his examination which I am very much pleased to hear of; I felt sure he would do well: and the examination, I understand, was a very hard one, which makes it the more gratifying.

The memoir of Cobbett, in Sir Henry Bulwer's Historical Characters, is very entertaining, and a very just estimate, I think, of the man's character. That of Canning, which I have not yet finished, is also very agreeably written. To Mackintosh, I think, he does scanty justice.

I am very glad you met the Arthur Herveys at Mary's,—they are delightful, especially Sarah, who is one of the most charming girls I have ever known.

We are both well, and it is a great comfort to see that Fanny has so very much recovered the power of walking.

I am looking forward with great delight to Charles and Mary's arrival.

Ever your very affectionate brother,
CHARLES J. F. BUNBURY.

JOURNAL.

February 14th.

Dear Charles and Mary Lyell arrived. I walked round the grounds with Charles. We had a very pleasant, quiet evening with them.

Fanny and Mary went to Mildenhall and returned to dinner. I had a pleasant walk with Charles Lyell.

Finished the Barton Estate accounts.

_____

Sunday, February 16th.

A remarkably beautiful day. Strolled about with Charles and Mary, and had much pleasant talk. I went alone to afternoon church and read in the evening Kingsley's excellent sermon on Discipline. I finished Motley. A most valuable and masterly history. These last two volumes (the 3rd and 4th) are perhaps the best of all. I have already mentioned, what to me has been the one drawback to the pleasure of reading them (and that not a very material one) the needless repetition of the praises of republican government. The negociations and intrigues which occupy a good deal of the latter half of the 4th vol. are tedious, but perhaps they could not with due regard to their importance at the time have been told much more shortly. This history ends with the conclusion of the Peace, nominally only a truce for 12 years, signed at Antwerp, April 1609: by which truce or treaty the Independence of the United Netherlands was acknowledged, and they were left virtually in possession of everything for which they had contended. The war had lasted —dating from the arrival of the Duke of Alva in the Netherlands—42 years. I doubt whether there is anything in history, ancient or modern, more glorious than the struggle maintained by the Dutch

for such a length of time, against a power which 1868.
seemed so overwhelming: nor are there many things
more satisfactory to my feelings than the result.
Holland came out of the struggle, not only
victorious, but one of the most wealthy and prosperous nations in Europe, while Spain was utterly
exhausted and effete. The wealth and flourishing
condition and real power of the Dutch Republic at
that time are well shown by Motley at the conclusion of his history, and they are indeed very
wonderful when we consider the scantiness of its
natural resources.

February 17th.

A beautiful day. Walked with Charles Lyell.

Dear Mrs. Kingsley and her two daughters Rose
and Mary arrived. We had a very pleasant stroll
together about the grounds.

February 18th.

We all went to Bury in the carriage, and saw St.
Mary's Church and the Abbey grounds. Charles
Lyell and I had a talk with Mr. Prigg.* Kingsley
arrived in the evening,—pleasant talk.

February 19th.

Rain. I looked over dried plants with Kingsley,
and after luncheon spent an hour very pleasantly,
showing minerals and fossils to the ladies.

* A shopman at Bury and a geologist.

1868.   The Abrahams arrived.   Patrick Blake and the
Bains dined with us.

February 20th.

Kingsley went early to Cambridge and returned in
the evening.   We went over to Hengrave.   Sir
Edward and Lady Gage showed us over the house
and were very good natured.

Friday, 21st February.

We had a large dinner party.   The Gages,
Wilsons, Chapmans, Lady Cullum, Hortons, &c.

February 22nd.

Confined to the house with a cold.   All the rest
of the party made an expedition to Hardwick.   I
wrote various botanical notes.   Read the Odyssey.
Sad news by telegraph, compelling Mrs. Kingsley to
leave us, thus breaking up our delightful party.

February 23rd.

Dear Mrs. Kingsley left us early, being obliged to
go to London to see her sister.   Read two of Kingsley's sermons, 6th and 7th of his new volume.

February 24th.

Charles and Mary left us.   I need not say that
I was sorry to part with them.   We have had a
delightful visit,—first, from the 14th to the 17th, the

Lyells alone; since the 17th, the Kingsleys also; and they seemed to suit one another, and to harmonize very well indeed. Charles and Mary are in very good health and spirits, and have been delightful. Mary's beauty in her 60th year is wonderful. I hardly need repeat, that I admire and value them both above most other people on earth.

Charles Lyell is just finishing the 2nd volume of his new edition of his "Principles." His thoughts are as usual much occupied with Geology and the sciences connected with it, and especially with the Darwinian theory, for which he has become quite an enthusiast. He is delighted with the success of Darwin's new book, which has sold much better than was expected, and he says that but for *his* encouragement, Darwin, who was very despondent about it, would hardly have brought it out, or at least not nearly so soon. Lyell does not allow himself much time to read other books than those bearing on Geology, but he is delighted with Motley, as I have been too.

Had a very pleasant walk with Kingsley. A visit from Lady Hoste and Mr. Brune,* her brother.

February 25th.

Beautiful day. Kingsley went early to Cambridge to his lecture, and returned in the afternoon. Mrs. Kingsley came back, to my great joy and surprise. I went on with my Memoir and with the herbarium catalogue;—a very pleasant, quiet evening.

---

* Mr. Trideaux Brune

February 26th.

Finished reading Sir Henry Bulwer's "Historical Characters;" an entertaining book.

Began to read Agassiz's "Journey in Brazil." Went on with my herbarium catalogue. Had a very pleasant walk about the grounds with Mrs. Kingsley and the girls.

February 27th.

We spent the evening looking over prints with the Kingsleys.

Friday, 28th.

Beautiful weather. Looked over dried plants with Kingsley. All the party except myself went to Stowlangtoft. Another delightful evening with the Kingsleys.

Saturday, 29th.

Very stormy. I spent the morning delightfully in shewing part of my Brazilian collection to the Kingsleys. The Louis Mallets arrived. Lady Cullum and Patrick Blake dined with us.

Sunday, 1st March.

Read Kingsley's Sermon on the "End of Religion." Afternoon very pleasantly spent in looking over prints with the Kingsleys. Kingsley read us a Sermon in the evening.

Looked over dried plants with Kingsley in the morning. Our dear friends the Kingsleys left us, to my great sorrow; it has been a great delight to have them and their two charming girls, Rose and Mary with us for a fortnight. I love and admire them both heartily, and I believe the regard and affection are reciprocal. Mrs. Kingsley appears to me quite as much superior to the average of people one meets with as her husband. One of Kingsley's most remarkable gifts is that versatility or mobility of mind which enables him to be (in the best sense), "all things to all men,"—to adapt himself to all sorts of society; to harmonize to a certain degree with all sorts of men, and always to find out and draw out whatever is best in them ; perhaps the great secret of this is an immense sympathy,—it is a gift in which I feel myself specially deficient.

Kingsley continues to take a keen interest in everything connected with Natural History, though he has but little time at his command for such studies. He told me of a plan he has,—which I earnestly urged him to carry out: of writing a Monograph on the Natural History (Geology included) of his own district—the plateau of the Bagshot sands on the Hartford Bridge flats. This district he has studied thoroughly, and he talked much to me of his theoretical views respecting its formation.

Kingsley said that the history of the corruptions in the Christian Church between the time of the Apostles and the establishment under Constantine

1868. is completely obscured. He believes that the various forms of Gnosticism and most of the early heresies grew out of the influence of Buddhism, which had pervaded a great part of Asia. Buddhism he says, was doubtless a reformation of Brahminism, getting rid of castes, and of the grossest superstitions.

He thinks that Motley and many other Protestant historians have overlooked the importance of the Anabaptist atrocities at Munster (in 1532). Those monstrous doings made a deep impression on men's minds: they contributed more than anything else to set the best men, such as Erasmus and Moore against the Reformation, and they had a great effect in instigating the Catholic sovereigns to persecution. Nor was this entirely without excuse, for the authorities might with some show of reason apprehend that similar effects might follow from the development of religious innovation in other districts.

Kingsley said that Swinburne, in his Chastelard, appears to have given a true view of Mary's character, and of the relations between her and Chastelard. He said that Swinburne's volume of objectionable poems, shows true poetic genius, and the most abominable of them all has the highest poetic merit.

Speaking of Froude in his Short Studies being rather hard upon Erasmus, he said it would have been well for Europe if the spirit of Erasmus had prevailed more in the Reformation.

Kingsley thinks that contrary to the common

belief, the American Indians were *not* a strong and healthy race at the time when the Europeans first came in contact with them, but were on the contrary a diseased and enfeebled race, and probably a dying-out race. This, he says, appears from the accounts of the early travellers in North America, French as well as English. He said, what is certain, that what the Ultra-radicals are now aiming at, is to bring about a confiscation of the landed property of Ireland, in order that they may make use of this measure as a precedent for the confiscation of landed property in England.

———

Wednesday, March 4th.

The Mallets went away early.

———

Friday, March 6th.

The Barton rent audit—very satisfactory. Read the Odyssey, book 5, verse 228 to 281, a very particular and minute description of the building of a boat by Ulysses, a difficult passage on account of the quantity of technical words in it, very well explained in Smith's "Dictionary of Antiquities,"— article on Ships.

———

Saturday, March 7th.

Read the Odyssey book 5, verse 281 to verse 381. A fine description of Neptune, exciting the great storm which destroyed Ulysses' vessel just as he had come in sight of the land of the Phænicians.

———

Sunday, March 8th.

A considerable fall of snow in the morning. But it soon melted, and the afternoon and evening very stormy and cold. I read prayers with Fanny, and read two of Kingsley's sermons.

* * *

Tuesday, March 10th.

Read Odyssey to end of book 5. The description of the miserable condition of Ulysses on his first reaching land, is exceedingly natural and vigorous.

* * *

Wednesday, March 11th.

A very rough, stormy, disagreeable day. We travelled in our own carriage to Stutton, to our dear friends, the Mills', and found them tolerably well in health, and very kind and cordial as ever. John Hervey staying in the house.

* * *

Thursday, March 12th.

A beautiful day. In the morning I strolled about the grounds of Stutton by myself. Then we looked over Mr. Mills' collection of Suffolk History and Topography. In the afternoon the ladies and I and John Hervey strolled about the grounds and the river bank. Captain Spencer arrived.

* * *

Friday, March 13th.

Mrs. Mills took us to Woolverstone to see the Bernerses. Mr. Berners showed us his hot-houses.

* * *

Saturday, March 14th.

We returned home.

LETTER.

Barton,
March 15th, 1868.

My Dear Lyell,

I am very curious to know what you think 1868. of Agassiz's "Brazil," or rather of his notions about the Geology of Brazil. I have very lately read his chapter 13 — "Physical History of the Amazon Valley," and am at a loss whether most to admire his ingenuity, and his power of observation and generalization, or to wonder at the extravagance of his hypothesis. I have no doubt of the value of his observations. I very well remember the bright, red clay which he describes so particularly, not only in the neighbourhood of Rio, but covering (or forming) the whole of the *campos* or undulating table-land in the southern part of the province of Menas, from Barbacena to Ouro Branco, or further. It is there cut by the rains into strangely deep ravines and gullies; and in that part between Barbacena and Ouro Branco) it contains a great quantity of angular sharp-edged fragments *(not pebbles)* of quartz and brown ironstone.

I find it specially noted in my Journal, that I did not see any which appeared in the least rolled or water-worn. I find also a note in my Journal, that in one place I observed several blocks of greenstone and of large grained granite, much weather-worn, standing out from the red clay. These, I suppose

1868. are some of Agassiz's "boulders." Agassiz points out very clearly and forcibly, what I remember struck me much in that country (smatterer as I was in geology), the extent to which all the rocks, granite and others, are decomposed. He acknowledges the difficulty of distinguishing, in many instances, between these decomposed rocks, and the deposits which he is pleased to call *drift*. Perhaps some may be profane enough to doubt whether he could always discriminate them so clearly as he professes. I well remember that when I was travelling there, I took up a kind of vague notion, I cannot call it a theory, that the red clay formation of Minas was a modification of granite or some such rock decomposed *in situ*. But of course, Agassiz's observations on the general extension of the same red clay over all kinds of rock, dispose of this notion.

It appears to me that Agassiz begs the whole question, in assuming that this red clay of Rio is glacial drift. If there ever was a time when the hills about Rio, down almost to the sea level (for red clay extends very far down) were covered with a thick crust of ice, *then* I admit, that at the same time, the valley of the Amazons (and, I should add, the whole earth) might be in the same way packed up in ice.

But surely, such an astounding doctrine requires a much stronger and broader basis of facts to rest upon than Agassiz has shown. I think he does not seem sufficiently awake to all the considerations involved in such a hypothesis.

It is clear that no plants or land animals, except 1868. those of truly Arctic or Antarctic types, could have lived through such a universal winter as he imagines, when Equatorial America was cased in ice ; therefore it is out of the questions that there should be any specific identity between the præ-glacial and post-glacial forms—or any generic identity, indeed except in the case of such genera as will bear a Polar climate. I do not know how far this may hold good with marine animals, nor how far this universal winter may be reconcileable with your Miocenes and Pliocenes ; but I think you will do well to look carefully to your Tertiary domains. I should be very curious to know more about those *fossil leaves* which Agassiz found in his glacial deposits on the banks of the Amazons. If they are found to agree with the recent flora, nothing would ever convince me that his glacial hypothesis is anything but a dream.

What do you say to his laminated clays, with a true *slaty cleavage* produced by the action of the sun? (p. 413). Is not this a startling novelty in geology?

For the rest, setting aside these glacial visions, the book is an excellent one, and most interesting to one who is interested as I am, in Brazil, and all that concerns it. All Mrs. Agassiz's part of the book is delightful. Her descriptions of all the country about Rio, are as truthful as possible, and therefore I feel no doubt that what she says of the Amazons and the rest of the country is equally so. The botanical observations are excellent, as far as they go, and those on fish, most curious and

1868. astonishing. I was much struck by Agassiz's lectures reported in the first chapter; they appear to me to be very valuable, giving just the sort of information and of advice, which would be the most useful preparation for naturalists about to enter upon such a country. If I had had any such hints to guide me, as to "what to observe," I might perhaps not have wasted my time in Brazil so much as I did.

One thing more I must say, what I think very important in the geology of this book is—that Agassiz has pointed out forcibly, the enormous extent and rapidity of sub-ærial denudation in tropical countries. I have no doubt it is much beyond what geologists in general would be apt to allow for.

I have read the article in the *Edinburgh* on "Liberal Education," and meant to have said something about it, but my letter is already so long that I must put off that subject to some other day.

Much love to Mary,

Ever yours affectionately,

CHARLES J. F. BUNBURY.

————

Barton,
March 16th, 1868.

Dearest Katharine,

I am indeed deeply grieved to hear of the serious illness of your dear boy Arthur, and you may be assured that I, as well as Fanny, feel most sincerely for your great anxiety and sorrow, indeed I think you will not doubt me ; you know well enough my affection for you and for your children. Most

earnestly do I hope and pray that it may please 1868.
God soon to relieve your anxiety, and to restore your
dear child to health. We shall be very anxious till
we hear decidedly favourable accounts of him.

We received this morning the first news of his
illness, in your letter to Joanna, which she forwarded
to us. It is some comfort, if he *was* to be ill, that it
has happened at Berlin, where you have the consola-
tion of Leonora's company and help, as well as a
good house, and all the medical and material
resources of a great capital. I entreat you most
earnestly to take care of yourself and not to risk
the making yourself ill by anxiety or over-watching.
Take care of yourself for the sake of all who love
you. I will not trouble you with a long letter under
these circumstances ; I have written these few
words only to assure you of our earnest sympathy
and affection.

I have pretty well shaken off the cold which hung
upon me so long, and am nursing myself up against
next week, when I have to receive and attend the
Judges of Assize, being Sheriff this year.

Pray give my love to Leonora and her dear
children.

Much love to dear Frank, who I am very happy
to hear is settled so much to his satisfaction as well
as yours.

Once more, I say, be assured of my earnest
sympathy in your trouble, and of my affection under
all circumstances.

Ever your very affectionate brother
CHARLES J. F. BUNBURY.

JOURNAL.

1868. Drove to Ickworth, and I saw the dear Arthur
Herveys, and spent some time with them. Charming as they always are.

————

Finished Agassiz's Journal in Brazil: an excellent and valuable book, and specially interesting to
me as the country is one which has always had a
special attraction for me. It is an account of the
very remarkable scientific expedition which Agassiz
with a whole staff of scientific assistants, (ardent
young men of science from New England) performed in Brazil, principally on the Amazon river
in 1865 and 1866. The narrative is written by Mrs.
Agassiz, and is delightful, excellently well written,
with only a very few Americanisms; clear, sensible,
with a natural unaffected eloquence at times, and
remarkably free from nonsense: giving a most
favourable impression, both of the heart and intellect
of the writer.

The illustrations, especially the views of the
vegetation, are by far the best that I have seen in
any book (of moderate size and cost) on tropical
countries, and are especially valuable as being taken
from photographs, and therefore thoroughly true to
nature. The portrait of a Palm on the frontispiece,
"the tree entwined by Sipos," (page 54) the two

views in the Botanic Garden of Rio; the view of 1868.
Botafogo Bay (p. 81) are as good as can be.

The party arrived at Rio in the latter part of
April, 1865, and remained in that neighbourhood
till towards the end of July, making an excursion
into the interior as far as the river Parahyba.
Remembering what travelling was, even in the
neighbourhood of Rio in my time (1833 and 1834),
it is surprising and amusing to me, almost tantalizing
too, to read of the facilities that Agassiz met with,
the good roads, the coaches and omnibuses, and
even a railroad (though this does not appear to be a
particularly safe one). Thence they went by sea to
Para, and up the Amazon in a steamer, which
was placed entirely at their disposal, so that they
could stop when and where they pleased.

They went up as far as Tabatinga, on the very
frontier of Brazil and Equador, making frequent
halts, and some for a considerable time, at places of
particular interest in a natural history view.

Agassiz's more especial scientific object was to
study the fishes of the Amazon and its tributaries,
of which comparatively little was known, and his
success was really astonishing. By his own
incessant exertions and those of his fellow travellers
and of very many amateur assistants (for in a
multitude of instances he found the Brazilians ready
and eager, and well able to help him) he collected
an almost inconceivable number of new species and
new forms of fish, and not only collected and
preserved them, but (as it appears) examined them
while fresh, and had drawings made of a great

1868. proportion of them. The wealth in fishes of that great river system, is as surprising as the multitude of the rivers themselves. It is not merely that you meet with new species from time to time as you ascend the river, but that different parts of the course of the Amazons, without any evident difference of local conditions, have entirely distinct *sets* of species, and the same with its tributaries. The number of *very small* fishes, which had been entirely overlooked by previous naturalists, appears to be enormously great,—not only the abundance of them, but the variety of species. One great peculiarty is the number of species of so-called *forest fishes*, which are never found in the open river, but only in the inundated parts of the forests.

During his stay at Para, Agassiz obtained more than 50 new species of fresh-water fishes. Early in his voyage up the river (between Tajapurá and Gurupá), he collected 20 species (15 of them new) in two days, and very soon after 84 species (of which 51 were new) in twelve hours. He might well say (page 164) that it was wonderful and that he could no longer put in order his new acquisitions as fast as they arrived. Afterwards (p. 273) it is mentioned, that in two days near Manaos (formerly Barra do Rio Negro) he obtained 70 new species. He mentions also, that in the course of the whole expedition, the draughtsman of the party made about *eight hundred* drawings of fishes.

But though this was his primary object, he does not seem to have neglected any branch of natural history, and all his notes on these subjects are 1868. valuable. He has not himself made any special study of botany, nor, as it appears, was any one of the party a botanist, yet his botanical remarks are always worth attending to. Particularly his observations on the replacement or representation of various families by others, in the tropical and temperate zones.

I have been much struck by his lectures, given on board ship to the members of his scientific staff, and reported in the first chapter. They are admirably well-suited to guide the researches of eager young naturalists in such a country. Such helps to observation would have been of infinite service to me when I went thither, and might perhaps have helped me to make a better use of my time.

In geology, Agassiz observed much in this expedition, and his observations *as far as they are strictly* observations are no doubt trustworthy and valuable. His description of the peculiar red clay, which covers all the rocks in the country about Rio de Janeiro, agrees perfectly with my recollections and notes, though his hypothesis as to its origin (that it is *glacial drift*) is very astonishing. His observations on the vast extent of this clay formation, and on the uniformity of the rocks throughout the immense valley of the Amazons, are exceedingly curious. His remarks on the extensive decomposition of the rocks in places by the action of weather, and on sub-ærial denudation, very good.

1868.   Agassiz's glacial speculations appear to me utterly extravagant, but I have neither time nor space to enter into that subject at present.   I have written largely upon it to Charles Lyell.

————

Friday, 20th March.

Sarah and Kate Hervey came to afternoon tea with us.

————

Saturday, 21st March.

A very fine day.   Strolled about the grounds with Fanny.   Towards evening weakness in my eyes came on.

————

Monday, 23rd March.

Mr. Lott arrived.   As I as Sheriff had appointed him my Chaplain, he and I went in my carriage to meet the Judges.   Half-way on the Bury road, a cavalcade of men on horseback (40 of my tenants)* met us, and escorted us to the railway station.

Baron Martin arrived, and the Commission opened; then we went to St. Mary's.   We had a good sermon from Mr. Lott.   Fanny went with us as far as Bury, and saw the procession from the Angel.

We had in the evening at dinner the Judges, Martin and Bramwell, and a large party.

————

* Mr. Phillips, who headed the procession, made a very kind and flattering address —(F. J. B.)

The Assizes concluded. Mr. Lott went away. I enjoyed a stroll about the grounds. We had a farewell visit from Sir John Walsham.

A weakness in my eyes in the first place, and then the assizes have occasioned a gap in my studies.

This day I resumed the Odyssey, and read the last 93 lines of the 6th book.

I also read the first chapter of the new volume of Lyell's "Principles," the 26th chapter of the book. This relates to Etna; the chief novelty in it is the very vivid and striking description of the changes produced in the Val del Bove by the eruption of 1852, and witnessed by Lyell in 1857 and 1858. All this chapter is excellent.

Went with Fanny to see the new foal, "Magdala."*

Clement arrived. We went to Bury, and had afternoon tea with Sir William and Lady Hoste.

Mr. and Mrs. Percy Smith arrived. We took them to see the Girls' School and the Vicarage. The Abrahams and Patrick Blake dined with us.

---

* He met with an accident and had to be shot in 1890.—(F. J. B.)

1868.   April 3rd.

The Percy Smiths and Clement went away early.
Captain Horton called.

* * *

April 4th.

Read chapter 36 of Lyell:—on Variation of
Animals and Plants, in the domesticated state,
and on the effects of selection by man. A very
interesting and remarkable chapter, full of curious
facts: avowedly derived in great measure from
Darwin's new book on "Variation," but very well
selected and arranged. I think, however, that it
would have been no loss at all, if the notice
of Darwin's hypothesis of *Pangenesis* had been
omitted. It had no essential connection with
Lyell's plan, he could not afford space to dwell upon
it sufficiently to make it even intelligible: and as
it stands in this chapter, it strikes one as merely a
wild fancy.

We dined with the Chapmans at Bury: met the
Bishop of Ely, the Arthur Herveys, the Wilsons,
Lady Cullum and Mr. Image.

* * *

Sunday, April 5th.

A most beautiful day. Went with Fanny to
morning church; afterwards we strolled together
about the grounds.

Read Dugald Stewart's* chapter 3, section 2, of
the "Affections of Kindred," and section 3 of
"Friendship." It is remarkable that in his section

* Dugald Stewart's "Active and Moral Powers."

on "Friendship," Dugald Stewart seems to assume 1868.
that it can exist only between persons of the
same sex, at least, he makes no allusion to the
possibility of a friendship between man and woman.
I believe this is quite a mistake. I believe that
most true and unalloyed friendship, as well as most
delightful, may exist between man and woman when
they are no longer young.

Monday, April 6th.

We went through the shrub. The Wood Anem-
ones in full beauty.

Wednesday, April 8th.

Henry (little Henry) arrived.

Thursday, April 9th.

We went with Henry to Ickworth, and spent the
afternoon with the Arthur Herveys, Lord Arthur
very busy. Pleasant talk with Sarah. Read
chapter 38 of Lyell on the "Geographical Distribution of Animals and Plants," and more especially
of Mammalia, with a particular exposition of the
provinces of zoological geography proposed by Mr.
Sclater, and of Mr. Wallace's explanations of the
anomalies:—excellent.

I object, however, to his ranking the tiger among
the Mammals of the "Palæarctica" region: I should
think it is certainly a tropical animal, which has
gradually extended its range far to the North.

1868.   I received a long and very interesting letter from
Charles Lyell, on Agassiz's "Brazil." I am very
glad to find that he objects as strongly as I do to
Agassiz's glacial fancies.

———

Saturday April 11th.

We spent the afternoon with Lady Cullum, at
Hardwick. Walked round the grounds. Saw the
beautiful bank of Forget-me-nots. Clement arrived.

———

April 12th.

Easter Sunday. Odiously cold weather. We went
to morning Church and received the Sacrament.
Read Dugald Stewart, chapter 3, section 4 of
"Patriotism" and section 5 of "Pity." Both very
good, especially the latter. Read Kingsley's
sermon on the "Lilies of the Field."

═══

LETTER.

Barton,
April 13th, 68.

My Dear Lyell,
        Very many thanks for your letter upon
Agassiz, which has interested me much, and I
am delighted that you are as much opposed as I
am to his hypothesis of burying Brazil under ice.
I shall now proceed at once to some remarks upon
the Darwinian chapters in your new volume so far

as I have read them,—that is, chapters 35 to 39. 1868
I am aware that you will be away from home, and
it will be some time before you receive this letter :
indeed, probably it will be some time before it is
finished: but I will begin writing now, because,
when written it will keep, and I can say what I have
to say, so much better with the pen than by word of
mouth. I must begin by saying that I think this
part of your book admirably clear and most
instructive. Every reader will see at once that you
are a zealous advocate of the Darwinian theory,
but I do not think you can be said to be bigoted,
and I much admire the candour with which you
acknowledge the difficulty with respect to *hybridity*
(especially at p. 321). This is, in fact, *the* great
difficulty in Darwin's way, and it does not appear to
me to be as yet at all overcome.

Now for my remarks, which in general apply
rather to opinions or facts which you bring forward
on the authority of others, than to what is strictly
your own.

First—I understand you (or rather Darwin) to say
that where a genus or other group includes a great
number of forms, running much in toone another,
with slight differences between them, this indicates a
comparatively *modern* group, "so that there has not
been time for the causes of extinction to make gaps
in the series of new varieties." (p. 340). Now, will
this always hold good ? We may confidently
affirm that the *Ferns* (including the Lycopodia) are
one of the most ancient families of plants now in
existence ; they are found well characterized in the

1868. most ancient deposits which contain any distinct traces of land plants. Yet there is no family of which the species run more into one another, or in which botanists are more puzzled to fix the limits of species and varieties.

2.—I doubt the correctness of Hooker's opinion (p. 305) that cultivated races of plants when they run wild, do not revert to the likeness of the original wild stock. One of the great difficulties in ascertaining the true native country of plants which are extensively cultivated (as you may see in many places in Alphonse de Candolle's "Geographical Botanique") consists in this, that it is often so difficult to determine whether individual plants of those kinds, which are found growing *apparently* wild are *really* wild or the relics of former culture. Would there be this difficulty, if cultivated races never lost their distinctive characters ?

Perhaps however Hooker would say, that these are not instances of real well-marked varieties, but of mere *variations* in luxuriance, like the cultivated states of the common red and white Clovers.

My other remarks refer to mere matters of detail. I much admire your exposition (chap. 38) of Sclater's " Regions of Zoological Geography," but—

3.—I must object to your naming the *Tiger* among the animals properly belonging to Northern Asia. I cannot doubt that his *specific* centre was in tropical Asia ; and that he gradually spread from thence to the north.

4.—Is it always true that our domestic animals are such as were social in their natural state ? (see

p. 302).   Is the wild cat a social animal ? or is the   1868.
jungle-fowl ? or is the rock-pigeon (which Darwin
admits to be the original of our tame pigeons) more
a social bird than the wood-pigeon ?

   5.—This is merely a question suggested by what
is said in p. 355, of the *swimming* power of quadrupeds).   Is any instance known of any of the
monkey kind being able to swim ?   I never remember to have read of such.

   What Wallace and Bates observed about the
range of various species of monkeys being limited by
the great South-American rivers, appears most
natural.   I should have been surprised if it had
been otherwise.   The small swimming power (if I
am not mistaken) appears to be one of the great
differences between the quadrumana and the *genus
homo.*

   As far as I have yet read, I think you have kept
pretty clear of the way that the younger Darwinians
run into, of representing the theory of natural
selection as having solved all the mysteries of
creation.   The   truth is, that the advocates of
Darwinianism or Lamarckianism, have a great
advantage, inasmuch as their's is really the only
theory (properly speaking) on the subject.   It is
necessary for clearness, to speak of the *theory* of
special creation, but the truth is, that *that* hypothesis
is merely negative ;   those who support it merely
mean to say, that species were created, they do
not know how, but independently of previous
species.   Evidently the advocates of a positive
theory have all the advantages of the *initiation*.   It

1868. remains to be seen what Agassiz will put forth, as he has decidedly thrown down the gauntlet to the Darwinians.

Do you remember our talking when you were last here, about Darwin's grandfather, the poet, and the scandal he gave by his theories? I have since been rather amused by hitting upon a passage in Davy's "Salmonia," concerning the "ingenious but somewhat unsound" speculations of Darwin (the poet) as to the hereditary transmission of peculiarities and the formation of new species thereby. He was in fact, a Lamarckian. So that Charles Darwin is himself an instance of the *hereditary transmission* of a propensity for daring theories.

I am sorry to find that you and Mary will not be in London when we arrive, but I hope we shall see you before long. What on earth are you going to do at *Southend?* in the mud of Essex.

Fanny is as well as usual; I caught cold at Mildenhall, standing about for two or three hours among excessively wet grass, in the plantations, on the dampest of days.

With much love to dear Mary,

Ever yours affectionately,

CHARLES J. F. BUNBURY.

JOURNAL.

Tuesday, April 14th.

Mrs. Hardcastle and her mother Lady Herschel and two younger sisters came to luncheon. I was very glad to see Lady Herschel again.

Occupied in putting the finishing touches to the Memoir of my father, which I began in August, 1866. I finished the first draft on November 26th, 1867, and since then have gone through it twice.

LETTER.

Barton,
April 15th, '68.

My Dear Katharine,

Enclosed I send you a brace of tickets for the Zoological Gardens for next Sunday. I was very glad to hear of your safe return home, and that dear Arthur had come back tolerably well: though I can well imagine that you will for some time feel anxious about him. You must miss your Frank very much, but it is a comfort to know that he is so well placed in all respects, and is happy. I am extremely glad to hear of Annie's recovery. As for us, we are in our usual state of health, and I am very thankful that I was not the worse for my cold service at the assizes. I do not think I ever enjoyed the spring flowers and the "vernal delight" more than this year; I have very great cause of thankfulness in finding my power of enjoying such things, now in my 60th year, entirely unimpaired. The spring blossoms, indeed, appear remarkably fine this year, and there has been no frost sufficient to do any mischief. I wish you could see our conservatory now, it is so lovely, so full of blossom.

1868.   I shrink from leaving home: though it will be a very great pleasure to see you and some few others in London, nevertheless it is a great effort to go.

Yesterday we had a visit from my old friend Lady Herschel, who came with her daughter Mrs. Hardcastle and two younger daughters to luncheon here. I had hardly seen Lady Herschel since the old Cape days, which we both remember with pleasure, and I was exceedingly glad to meet her again, and to talk of the Cape. She was very pleasant, and so was Mrs. Hardcastle.

Little Henry is spending his holidays with us: he has a remarkable passion for chemistry, is quite devoted to it, and delighted to spend his days in trying experiments: he is intelligent in other ways too, though far from fond of regular study, indeed as a school-boy of 13, he would be a phenomenon if he were fond of *that*. Clement is trying for a scholarship at Trinity.

I am deep in the Darwinian chapters of Lyell's new volume, and very much interested, and noting down criticisms or rather queries to send him. I have no doubt that when you are enough settled, to have leisure for reading, you will be delighted with this volume.

Fanny has all but finished her great catalogue, and my Memoir is very nearly ready for the press.

Ever your very affectionate brother,
CHARLES J. F. BUNBURY.

JOURNAL.

We with Henry went over to Mildenhall.   1868.

Visited poor old Mrs. Bucke. Mr. Lott came to luncheon with us: very pleasant as he usually is. Visit from Mr. Gedge, the curate, and his wife. We returned home to Barton: found Clement arrived. We heard of the death of General Simpson.

Wet weather. Read Kingsley's excellent sermon on the Good Samaritan. Read part of Acts xvi. in the Greek, with Alford's Commentary.

Furious rain and wind. Arranged the fossils I had brought from Mildenhall. Our dear Sarah and Kate Hervey came, and spent the afternoon looking over prints with us.

Arthur Hervey spent the afternoon with us : delightful as he always is.

Paid money to Mr. Sparke on account of sheriff's

1868. expenses.   Mr. Milbank and Lady Emma Osborne came to luncheon.

Mr. Fuller Maitland came to luncheon and to see our pictures.   We spent the afternoon in planting a variety of flowers in the Vicarage Grove and Arboretum.

Making our preparations for leaving home.   News of the great success in Abyssinia.   Finished a second reading of chapter 41 of Lyell's "Principles," on Insular Faunas and Floras.   He takes especially those of the Atlantic Islands, — the Madeiras, Canaries and Azores, — the groups with which he is best acquainted, and begins with a clear sketch of their geological characters: maintaining and apparently proving that they had been built up by volcanic accumulation, and are not remnants of a "Miocene" or Atlantic "Continent."   He then proceeds to notice the principal classes of their animal inhabitants, touching rather slightly on their floras: arguing that all the peculiarities of distribution tell in favour of the theory of variation. He remarks, very truly, that the general absence of indigenous Mammalia from small oceanic islands, even from those so fertile as Madeira and Teneriffe, goes far to prove that animals have not been placed by special creations in every locality that was suitable for them.

Next, as to Birds, he shows that those of the islands in question, are (with very few exceptions) the same as those of the nearest continents, Europe and Africa:—that there is little difference, in this respect, between the different groups, and none between the different islands of each group. This is very easy to understand, considering the flying powers of birds. But then he proceeds to compare the case of the birds of the Galapagos Islands : and here the contrast is certainly very strange. Although the Galapagos are not more than half as far from S. America as the Azores from Europe, yet almost all their land-birds are of species quite peculiar to them; and what is more, several of the species are confined to single islands.

I do not think that this extraordinary peculiarity of the Galapagos is sufficiently accounted for. The insects of the Atlantic Islands are in a very different case from the birds: a great proportion of the species and many genera are peculiar: and this is explained by their lesser powers of flight, particularly the case of the Coleoptera. But the land shells are the most remarkable of all in their distribution. Not only are almost all the species distinct from those found in Europe or Africa, but Madeira and Porto Santo, only 30 miles apart have very few species in common: and almost every islet has some one or more confined to it. Lyell contrasts this with the case of the British Islands, of which the land shells are almost all identical with those of the Continent, and where none of the smaller islands have peculiar species or even races.

1868. The reason of this is, as he shows, that all parts of the British group have been united together, and with the Continent, within very recent geological times: whereas the Madeiras have been separate ever since the Miocene Age. But he fairly admits that, seeing how unable the land shells have shown themselves to cross even such small barriers of sea as those between Madeira and Porto Santo and the Desertas, it is very hard to explain how the progenitors of the present species ever reached those islands from the Continent. He seems almost to acknowledge that for the present, this difficulty is insoluble.

Tuesday, April 28th.

Up to London. Henry returned to his school.

Wednesday, April 29th.

Wrote to Henry. Also on business to Messrs. Spottiswoode, the printers. Called on Edward and saw him. Then on Minnie Boileau and saw her. Bought a good copy of Williamson's "Oriental Field Sports," an old book which delighted me when I was a boy. It has fallen off surprisingly in price; this copy, a large paper one, handsomely bound, in very good condition, and altogether a very handsome book, cost me only 35s.

Bentham's evening party or Linnean *gathering* at the rooms in Burlington House. As last year, a superb display of rare, beautiful and curious exotics, supplied from Kew and from some of the principal nursery gardens; one of the most remarkable, though not one of the most beautiful, a fine specimen of the very rare Monizia edulis from Madeira—an umbelliferous plant which looks as if it meant to imitate a tree-fern; having a bare, upright, undivided, columnar stem, some six or seven feet high, crowned with a large tuft of handsome, deep green, curling, and much divided leaves. Some withered leaves hanging down and partly concealing the trunk, give it still more the look of a tree-fern. The specimen was not in flower or fruit.

Some beautiful ferns, especially a very fine (new ?) Aneimia of the Phyllitidis group; lovely Orchids, and superb specimens of the Anthurium Scherzerianum, with its brilliant scarlet spathes.

A great variety of curiosities, both of nature and art, exhibited on the tables and on the walls.

A fine collection of dried plants of the Bignonia family, with their fruits, sent by a Brazilian gentleman.

Bentham remarked to me, how much alike the flowers are in most of that family, while the fruits are very various, and many of them very curious and strange.

———

Thursday, April 30th.

Visit from Katharine. Went to the portrait exhibition at South Kensington, and spent nearly two hours there. It is a rich collection, less interesting on the whole than that of last year, but interesting in a different way, as including the portraits of so many whom we ourselves have known, or who at least have flourished in our time, as well as

1868. many more who were passing off the scene in our
youth ; here are the originals of numberless
portraits with which one is perfectly familiar from
the prints ; those of Cowper, Scott, Byron, Moore,
Macintosh, Romilly, Canning, young Lambton, &c.,
&c.

In point of art, this exhibition will not bear comparison with that of last year's and this is especially
true of the female portraits.   The ladies of George
IV. and William IV.'s time compare to little
advantage to those who were painted by Reynolds
and Gainsborough ; but no doubt the inferiority
must be imputed more to the artists than to the
ladies.

Met Sir Frederick Grey at the Gallery, and Mr.
Gurdon.

Looked into the South Kensington Museum.
Sally and her girls and Edward dined with us.

Friday, May 1st.

Mr. Geo. Spottiswoode came to me at 10 o'clock
a.m., and took charge of the MS. of the " Memoir
of my father."

Susan and Joanna came to luncheon with us.   I
called on the Moores and the Adairs.   Dear Minnie
and Sarah dined with us.

Sunday, May 3rd.

A splendid day.   Read a beautiful sermon in
Kingsley's last volume on " The Temple of
Wisdom."   We went to Wimbledon and spent the
afternoon with the William Napiers, very pleasantly.

Saw Charles Austen, Matthew Arnold and others, at the Athenæum. Susan dined and spent the evening with us.

———

Tuesday, May 5th.

We went to Norwood and spent the afternoon with Susan and Joanna. Joanna walked through the Crystal Palace and gardens with me.

———

Wednesday, May 6th.

Went with Fanny to an afternoon party at Lady Rayleigh's. Met Charles and Mary Lyell, John Herbert, Mrs. Ellice and Helen.

———

Thursday, May 7th.

William Napier came to luncheon, and went with us to the Royal Academy. A great crowd. Millais' picture of two old Grenwich Pensioners visiting the Tomb of Nelson.· A noble picture. Watts's portrait of Panizzi; an uncommonly vigorous and powerful portrait, worthy of the good old times. Two or three lovely children by Eddis and Sant.

———

Friday, May 8th.

Again very hot. We had a great many pleasant afternoon visitors: the Charles Lyells, Norah Bruce and Pamela Miles, the Bowyers, Colonel Kinloch and his daughter and sons.

———

Saturday, May 9th.

We went down on a visit to Sir Frederick and
Lady Grey, to their house at Sunningdale. A
young and very pretty Miss Grey* came to dinner.

———————

Monday, 11th May.

We returned from a very agreeable visit of two
days to Sir Frederick and Lady Grey. The
situation of their house is most agreeable. They
took us out, driving, through part of Windsor Park:
a most delightful drive, quite new to me. I was
charmed with the beauty of the woodland scenery,
the fine Thorn trees covered with blossom, the
glorious old trees, Oaks and Beeches, the picturesque
variety of ground, and the grand views of Windsor
Castle in the distance. It is, indeed, a palace and
a domain, worthy of a great sovereign.

Sir Frederick Grey's house is very well situated,
on a high ground, commanding on the south and
south-east, a very fine and extensive view as far as
the North Downs near Guildford, and even as far
as Leith Hill. Through a telescope we saw very
distinctly the grand stand at Epsom, and the tower
on Leith Hill. The situation is, indeed, extremely
good, for, as he pointed out to me, the ground
descends from it on all sides, so that they can never
be *built up*. The house has been built entirely
from their designs, and is admirable in its arrangements. I never saw anything more perfect. It is
on the Bagshot Sands,—Upper Bagshot I con-

* Afterwards Countess of Home.

ceive; the beds near the surface are gravels and sands, brown and yellow, like those of Sandhurst and Eversley; and at a certain depth is a thick bed of greenish clay, very retentive of water. This, I presume, corresponds to the green clay of the middle Bagshot, which Kingsley pointed out to me at Eversley. Sir Frederick showed it to me in a railway cutting close to his house. Below this, he said, is a fine white sand. In sinking wells, they obtain good water from the beds immediately over the clay; but a neighbour of his thought fit to penetrate through the clay, and found that he could reach no water without going to a great depth below.

From hence stretches far away to the south and south-west, the fine open heath country of Bagshot and Sandhurst and Eversley, and the Hartford bridge flats to the foot of the chalk downs.

In the drive through Windsor Park, I saw, for the first time in my life, red Deer at liberty (though of course not wild): I had never before seen them except in menageries.

Received the news of the death of Lord Brougham. I received from Spottiswoode the first proof sheet of my Memoir.

Tuesday, 12th May.

Went with Fanny, Minnie and Sarah to the Drawing Room:—a dreadful crowd—the whole day lost. The Bowyers and George Napier dined with us.

Wednesday, 13th May.

Read the Odyssey, book 7, verse 78-121: the description of the palace and garden of Alcinous. These were undoubtedly intended to be on the greatest scale of magnificence which the imagination of that age could conceive, but it is rather amusing to observe that the extent of the garden (or orchard) is only four acres.

Went to the Zoological Gardens. Spent some time there with much satisfaction: then went to see dear Katharine, and had a pleasant talk with her.

Thursday, 14th May.

Read Mr. Coleridge's fine speech on the Abolition of Tests in the Universities. Beautiful weather. Mary and Katharine came to afternoon tea. Minnie and Sarah and Edward dined with us.

Friday, 15th May.

Very fine and warm. Went out with Fanny to a reading at Madame de Bunsen's. Called on the Charles Lyells.

Saturday, May 16th.

Went with Fanny and Minnie to the Portrait Exhibition, which I had also visited alone on the 14th, so that I have now seen it pretty well. It is a very interesting historical gallery, though much less rich in fine works of art than that of last year. There are, however, some excellent pictures in the

"Supplementary" department: a few *Vandycks* 1868. and several *Reynoldses* and *Gainsboroughs*. In this part of the collection I noticed also a portrait of the famous Miss Chudleigh (so often celebrated and laughed at by Horace Walpole). I had often wished to see a likeness of her, and this is a very attractive one. It is a "pastel," by *Cotes*:—quite a Louis XV. beauty, extremely pretty and very saucy looking; but on seeing her again, I did not think her quite as pretty: but it is quite a "Boucher" face—a Sevres China nymph.

In the more modern part of the collection, I observed some excellent portraits by *Raeburn*, particularly that of Professor George Joseph Bell, but I am disappointed by his "Sir Walter Scott:" the head is undignified. The small portrait of Scott by *Leslie*, is much more satisfactory to me. The group of Scott and his friends by *Faed* is interesting. A vigorous portrait of Professor John Wilson by *Sir J. Watson Gordon*. "Charles Lamb and his Sister," by *F. S. Carey*—interesting. It is curious, remembering that Quakers were included in Lamb's "Imperfect Sympathies," to see how like an old Quaker he looks in this picture. Southey's face is unsatisfactory, and Coleridge, by *T. Phillips*, still more so: Coleridge looks as little like either a poet or a philosopher as I can well imagine; he looks thoroughly "Epicuri de grege porcus,"—fat and smooth and sleek, self-satisfied and comfortable. Lady Hamilton as a Magdalen, by *Romney*—most lovely. Mrs. Siddons, also by *Romney*, very beautiful.

1868.   This exhibition, together with what I have seen at different times at the British Institution, gives me a high admiration for *Romney* as a portrait painter.

Margaret Woffington, by *Hogarth* (from Lord Lansdowne) remarkaby beautiful and fascinating. Lady Warwick, with her Children, by *Romney*— beautiful.

We dined with the Bruces. Met Sir Cecil and Lady Beadon, Sir Bartle Frere, Sir George Young and others. Some good conversation.

Sunday, May 17th.

We drove out to that lovely spot Combhurst, and spent an hour very pleasantly with Mr. and Mrs. Sam Smith ; then to Wimbledon and dined with the William Napiers.

Monday, May 18th.

Went a second time to the Royal Academy, and saw it much better than the first time.

I noticed and approved *Maclise's* "Madeline after Prayer," "The British Captive in the Arena," by *W. V. Herbert;* *Cooke's* "Scheveling Trawler." "The Moonlit Shore," by *Teniswood."* "A Calm on the Humber," by *Redmore;* very like *Vandervelde.*

Leighton's "Acme and Septimus," classical and sweet, and less cold than most of his works. The same artist's "Actæa," a naked nymph reclining on the beach, much more natural than his "Venus," (last year's) but I do not at all like his "Ariadne." It is rather startling to see "Ariadne" represented as lying dead ; accustomed as we are to the myth 1868. which makes her become the bride of Bacchus. But I was told that the story which Leighton has followed, is told by Homer, and I find this to a certain degree true.

In the 11th book of the Odyssey, Ulysses sees Ariadne among other heroines, in Hades, and tells how Theseus carries her off from Crete towards Athens, but they did not reach it, for Artemis slew her in the island of Die, " on the information of Dionysis."

These last words are explained by Cowper in a note (I suppose on the authority of a Scholiast) to mean that Dionysis accused Theseus and Ariadne to Artemis, of violating the sanctity of a temple. But Homer does not even hint at any desertion by Theseus.

Our dinner party. Charles and Mary, Colonel and Mrs. Yorke, the Miss Richardsons, George and William Napier.

--- --- ---

Tuesday, 19th May.

Extremely hot. Susan Horner came to luncheon with us, and went out driving with Fanny.

I went to the National Gallery and saw it comfortably ; and afterwards to the Museum of Practical Geology. Called on Julia Moore and saw her.

--- --- ---

Wednesday, 20th May.

Called on Norah Bruce, and met William and Emily Napier there.

--- --- ---

Thursday, 21st May.

We dined at Sir John Hanmer's. Met Sir
Thomas Gladstone, Admiral and Mrs. Eden, Mrs.
Chetwynd, &c.

―――

LETTER.

48, Eaton Place, S.W.
May 21st, 1868

My dear Henry,

I am sorry to hear you have given up the
plan of coming to town this Whitsuntide ; I should
have liked to have a talk with you about North
Wales and various matters. I was much interested
by your letter, your account of your visit to the
Bishop, and the tour afterwards ; and thank you
very much for it. It is a long time since I
saw that part of the country (in June, 1842)
but I have a vivid remembrance of the pass
leading up to Llyn Ogwen, and the valley from
thence up to Capel Cerrig ; and the scenery about
Llanberis. Does the railway go near Tan-y-
Bwlch? I suppose not. I remember it is a
charming country along that road going from
Harlech to Pont Aberglaslyn and Beddgelert.

The shores of the Menai Strait, I remember are
beautiful, and the contrast between the two sides
very interesting. But after all, I do not know that
there is anything in Wales superior to the valley in
which you live.

We are going on pretty comfortable in London,

only Fanny has had a succession of ailments which 1868.
have disappointed her of a good deal of society.

In the first batch of hot days after we came to
town, she caught cold, and was debarred from going
to various parties ; and in the extraordinary heat of
the day before yesterday, her head suffered. However, she was able to go to Miss Coutts's and to
Lady Nugent's ball with Minnie and Sarah, and
seems all the better to day.

I am busy correcting the proofs of my Memoir of
our Father. Three sheets (of sixteen pages) are
now printed, and I hope to send you a copy in the
course of next month.

I have seen the Portrait Exhibition well ; it is not
quite so interesting as that of last year.*———

JOURNAL.

Friday, 22nd May.

Read the Odyssey to the end of the 7th book.
In this book, the arrival and first reception of
Ulysses, as yet unknown, at the palace of Alcinous,
is interesting throughout ; full of the peculiar,
primitive, simple dignity and generosity of those
heroic times.

Dear Sarah Hervey arrived to stay with us. We
dined with the Hutchings. Afterwards went to Miss
Coutts's—a great crush.

---

* End of letter lost.

Saturday, 23rd May.

Minnie, Sarah, Helen Ellice, John Herbert and Edward dined with us.

————

Snnday, 24th May.

Wet.   We went to morning Church with Sarah Hervey.

————

Monday, 25th May.

Went to the Linnean Society and heard Bentham's anniversary address—a review of the principal publications on " Biology " (Zoology and Botany) within the year.   It was characterized, like his other writings, by his good sense and clearness of intellect; by the power which he has, I think, in a remarkable degree, of penetrating to the heart of a subject and throwing away the rubbish of it.

We had at our dinner to-day—Sir John and Miss Kennaway, Mrs. Andrews, Mrs. Gibson Craig, Minnie and Sarah, Sir George Young, Douglas Galton, George Napier and Edward.

————

Tuesday, 26th May.

Visit from the Thornhills.   We went (Sarah Hervey, Fanny and I) to the Portrait Gallery. Stayed two hours there and saw it very pleasantly. A peculiarly interesting picture, which I had over-looked before, is the portrait of Cowper's Mother, lent by Mr. Donne,—interesting, neither for the beauty of the subject, nor for any special merit in the picture itself, but because it is the identical

picture on which Cowper wrote his famous lines— 1868.
those sadly beautiful lines. The portrait of Cowper
himself by *Romney* (well-known from engravings),
struck me forcibly, by the very strong, and (as it
appears to me) unmistakable expression of *madness*
in the face. Southey's remarks upon it appear to
me perfectly just. See life of Cowper, vol iii, p. 87.
Mr. Donne, whom I afterwards met at the Lyell's
said (on my speaking of this look of madness in
this portrait of Cowper) "Yes, it is the portrait of a
madman, painted by a madman." He then told
me (what I did not know before) that Romney had
a tendency to melancholy madness like that of
Cowper; only that in him (Romney) it did not take
a decidedly religious form.

Romney's Mrs. Siddons is certainly one of the
most beautiful faces in the whole Gallery: perhaps
the most beautiful—merely a sketch, but one hardly
wishes it more finished.

May 27th. Wednesday.

We went to the flower show at the Botanic
Gardens in the Regent's Park:—a very beautiful
show of fine and rare flowers, Ferns, &c., in vast
variety but I did not observe anything very remarkable which I had not seen at former shows.

In the great conservatory, a fine species of
Fourcroya in flower. Afterwards called on Mrs.
Berners, then went to Mr. Horton's, Grosvenor
Place, and stayed to see the parties return from the
Derby. It was a beautiful day.

Thursday, 28th May.

Splendid weather. Letters from Scott. Visits from William Napier and Mr. Harness. Went to the Horticultural Society's Gardens at South Kensington; the show of Rhododendrons and Azaleas from Waterer's garden is exceedingly beautiful. Afterwards went to the Royal Academy.

Friday, 29th May.

A visit from Lord Augustus Hervey. In the midst of it a great thunderstorm. Called on Lady Smith and Sir John Bell—both pleasant. We dined with the Thornhills.

Saturday, 30th May.

Beautiful weather. The 24th anniversary of our happy wedding day,—thanks be to God. I went to the levee.

Sunday, May 31st.

Went to morning church with Sarah Hervey.

Monday, 1st June.

Beautiful weather. Lady Rayleigh and Miss Strutt came to luncheon. Fanny's afternoon tea party—many pleasant people. We dined with the Charles Lyells:—met Sir William and Lady Codrington, Sir Henry and Lady Rich, Lord Rosse, &c.

Went to Boone's and bought a fine edition of the *Spectator*. A visit from Shafto Adair.

Wednesday, 3rd June.

Dear Sarah Hervey left us. We went to the flower show at the Horticultural Gardens, and I observed there a great rarity,—that very distinct and peculiar Fern *Actino pteris radiata*, the first time I have ever seen it alive, and I suppose the first time it has been raised in Europe. There were several plants of it, small, but looking very healthy and flourishing, under glass. Its. general likeness on a minute scale, to a Fan Palm, is very remarkable. The colour is a fine, rich, full bright green.

Thursday, 4th June.

Visits from Sir John Kennaway and Minnie. We dined with the William Napiers at Wimbledon. Met Lady Napier* (whom I was especially glad to see), Norah, Catty, Emily and George.

Friday, 5th June.

Catty and William Napier came to luncheon. I visited the South Kensington Museum, the Portrait Gallery and the Horticultural Gardens. In this second visit to these Gardens, I observed another novelty, *Davallia parvula*, a most exquisite little Fern, looking at first sight like a delicate Moss of a rich green colour.

* Widow of Sir George Napier.

1868.   There were also specimens of the Dionaea and
some beautiful specimens of Nepenthes.

——————

We spent the day with Susan and Joanna at
Norwood.   I rambled about the Crystal Palace
while the rest were at the Concert, and was well
enough entertained.   Mrs. Young and her daughters
joined the party, and afterwards Edward.   Susan
returned with us to Eaton Place.

——————

Susan very pleasant.   Fanny and I went to
afternoon church.

——————

Read the Odyssey, book 8.   The episode
of Venus and Mars is amusing; the description of the Phæacian dancing and ball-playing
—curious.   Susan left us.   Lady Napier came to
luncheon.   I accompanied her to call on Sir John
Bell.   Our dinner party consisted of Bentham,
Fanny Mallet, Mr. and Miss Donne, Erasmus
Darwin, Babbage, &c.

——————

Fanny went to Wimbledon, to be present at
Susy Napier's confirmation.   I visited the Charles
Lyells.

——————

Read the Odyssey, book 8, the end of the

book. The description (in a simile) of the woman 1868.
mourning over the body of her husband who has
been killed in defending his country, and of the
victors driving her away into slavery, is very fine
and very touching.

The idea of the Phæacian ships which guide
themselves without rudder or pilot, and know of
themselves whither their masters want to go, is as
wild as any thing in *Sindbad's* voyages.

We went to see a very interesting collection of
Marc Antonio's engravings, exhibited by the
Burlington Fine Arts Club. The ticket of admis-
sion was given me by Mr. Chetwynd, whom I had
met at dinner at Sir John Hanmer's on the 21st
May.

At dinner at Mrs. Young's, I sat by Mrs.
Walrond, a niece of Mr. Kingsley, (she was Miss
Grenfell), and found her quite charming—quite
worthy of her aunt and of her cousin Rose. She
is moreover very pretty.

————

Thursday, 11th June.

A long visit from Emily Napier* (not Emily
William).

Dear Sarah Hervey arrived. Our dinner party
consisted of Sir Henry and Lady Rich, Mr. and
Mrs. Brookfield, Sir John Kennaway, Norah Bruce,
Emily Napier, Cissy Ellice.

————

Friday, 12th June.

Joanna Horner spent the afternoon with us. I
drove out with her, Sarah Hervey and Fanny.

* His first Cousin, daughter of Sir William Napier.

1868. We went to afternoon parties at Mrs. Octavius Smith's and Mrs. Milner Gibson's. Called on the Boileaus and saw Theresa.

——— ——

Sunday, 14th June.

Splendid weather. Very hot. Went to morning Church with Fanny. Visit from Sir Henry Rich.

——— ——

Monday, 15th June.

Read the Odyssey, book 9. Ulysses makes himself known to the Phæacians, and begins the story of his adventures. What he says about the geographical position of Ithaca, is hard to understand, or rather to reconcile with facts. Edward concludes from this passage, that the poet of the Odssey, whoever he might be, was not personally familiar with the Island of Ithaca.

Went with Sarah Hervey and Fanny to a musical evening party at Lady Codrington's.

——— ——— —

Tuesday, 16th June.

Went out with Fanny and our dear Sarah Hervey to see Mr. Baring's beautiful collection of pictures, which he so liberally allows to be seen; I had visited it two or three times before in former years, but it is well worth seeing any number of times.

Afterwards we went to see Holman Hunt's new picture, "Isabella with the Pot of Basil," which is exhibited in King street; the subject (from

Boccacio is strange and unpleasant; the woman's 1868. face ugly, and the first impression generally rather repulsive; but the painting, the perfect reality, exquisite finish and richness of colouring—of the whole picture is wonderful.

A visit from Mrs. Walrond.

We had at dinner—Lord Dunlo, Lady Adeliza, Sir John and Lady Hanmer, Mr. and Mrs. Thornhill, Admiral and Mrs. Eden, Mr. and Mrs. Hutchings, &c.

Wednesday, June 17th.

Fanny, Sarah Hervey, and I went with Minnie and Sarah, Sir George Seymour, Colonel Seymour and others, to see Lord Hertford's pictures, in Manchester square. They are very seldom allowed to be seen, but Colonel Seymour had got leave for Minnie and her friends to see them, and she kindly asked us. They are certainly a splendid collection. The two that have left the strongest and most vivid impression on my memory, are—*Rubens's* famous "Rainbow Landscape," on which the effect of distance is perfectly wonderful; and a portrait of Nelly O'Brien, by *Sir Joshua;* a sitting figure nearly whole length, directly facing us. This is certainly one of the most lovely and admirable portraits by *Reynolds* that I have ever seen.

Our dear Sarah Hervey left us. She is succeeded by her sister Kate, who is a very charming girl, but Sarah is still my especial favourite.

The " Nelly O'Brien " at Lord Hertford's is

1868. certainly one of the most beautiful female portraits that I have ever seen. The face itself is not of extraordinary beauty; but the perfect, easy grace of the attitude, the natural, and at the same time, refined air of the whole figure, and the management of the light and shade, are perfectly charming.

We dined to day at Lady Napier's, and met Mr. Waldegrave and Lady Rothes, &c.

Thursday, 18th June.

Went with Fanny and Kate Hervey to the Royal Academy. A visit from my nephew Cecil. Sarah Hervey dined with us.

Friday, 19th June.

Fanny and Kate Hervey went to the Handel Festival at the Crystal Palace.

Read the Odyssey, book 9. The description of the Island lying of the land of the Cyclops, is beautiful, but (as I think Humboldt remarks—of most of the descriptions of natural scenery in Greek poetry) there is a certain *utilitarian* tendency in it; I mean a tendency to dwell specially on the fertility and productiveness of the land with a view to human wants and uses.

I had a visit from Augusta Freeman.

Saturday, 20th June.

We and Kate Hervey visited the Bishop of London, and Mrs. Tait at Fulham. The grounds of the Palace are beautiful, the freshness and verdure of the trees and grass, even in this dry weather, 1868. might make one suppose one's self far from London; and some of the trees are very fine.   There are a few noble old exotic trees still remaining, of those planted by Bishop Compton ; in particular, the finest black Walnut that I have ever seen, the oldest, I should suppose, in this country, and probably in Europe ; a gigantic Ilex, sadly shattered in a snow storm, two or three years ago ; and a Cork tree of uncommon size, though likewise a wreck.

The Taits, who were very kind and cordial, showed us besides the beautiful grounds, the new chapel and hall.

Fanny went out after dinner to see Mary.

Sunday, 21st June.

We went to morning Church.   I called on the Charles Lyells, and had a chat there ; also on the Berners—very pleasant and cordial.

Monday, 22nd June.

A visit from poor Sir John Boileau, sadly broken. Fanny and Kate went to Wimbledon.   I visited the Portrait Gallery and the Horticultural Gardens. The Frederick Freemans dined with us.

Tuesday, 23rd June.

Read chapters 44 and 45 of the new edition of Lyell's "Principles."   I had read before the end of April, as far as chapter 42 ; but finding chapter 43 (on the origin and distribution of man) too hard to

1868. read midst the distractions of London, I laid the book aside.   I have now taken it up again, beginning with chapter 44, intending to reserve the difficult 43rd chapter till I am quiet in the country.

Went to Regent's Park Road, and had a pleasant talk with Katharine.   Minnie, Sarah and George Napier dined with us.

――――

Read the Odyssey, book 9.   Part of the adventure in the cave of Polyphemus.   There seems to be some probability (See Lane's Notes to the Story of Sindbad in the Arabian Nights), that this story is of Eastern origin.

Read chapter 46 of Lyell's "Principles."   The principal novelty in this edition is the description of the submarine forest at Bournemouth.

Went with Kate Hervey through part of the Portrait Gallery.   Sarah Hervey, George and William Napier dined with us.   Fanny went at night with Kate to two balls.

――――

Read chapter 47 of Lyell's "Principles."   The novelty in this is a very clear and useful summary of the recent discoveries and speculations concerning pre-historic man, especially concerning the Neolithic and Palæolithic ages.   As to the so-called periods of Bronze and Iron, I have doubts as to their distinctness—unless in some particular countries of limited extent.

We dined with the Adairs. Met Mr. and Lady 1868. Gertrude Foljambe, Sir George and Lady Nugent, Mr. and Mrs. Micklethwaite, &c.

June 26th Friday.

We received the news of the death of poor old Mrs. Bucke.

I corrected the 15th sheet of my book, containing the end of my own work,—that is, of my Memoir of my father; and containing also his history of the family of Bunbury, which he had intended to have privately printed. Five or six more sheets, I suppose, will complete the work.

Saturday, June 27th.

A party at luncheon:—Mrs. and Miss Moore Smyth, Minnie and Sarah and George Napier and Joanna Horner. We dined with Sir Henry and Lady Rich.

Sunday, 28th June.

We went to church before breakfast and received the Sacrament: afterwards to morning service and heard a good sermon from the Bishop of Oxford, (Wilberforce.)

Visited Mrs. Walrond. Had a delightful talk with her and found her quite as charming as the first time.

Monday, 29th June.

A visit from Arthur Hervey, whom I was delighted to see. We went to Veitch's nursery garden and

1868. ordered some plants. Dear Sarah Hervey dined with us.

_____

Tuesday, 30th June.

Arthur Hervey and Henry Bruce breakfasted with us—very agreeable. We dined with the Bishop of London at Fulham.

_____

July 1st.

Our dinner party consisted of Mr. and Lady Gertrude Foljambe, Mr. and Mrs. Walrond, the Charles Lyells, the Henry Bruces, Mr. Clark, Lady Bell, William Napier, Edward. Much delightful talk with Mrs Walrond.

_____

Thursday, 2nd July.

Arthur Hervey came to luncheon with us. Edward dined with us, and Minnie and Sarah came in the evening.

_____

Friday, 3rd July.

Packing up. Went to an afternoon party at Katharine's in celebration of Rosamond's birthday.

_____

Saturday, 4th July.

A bitterly cold and boisterous north wind. We travelled post down to Barton. Left Eaton Place at 9.40, and reached Barton at 7.5. All safe and well at home, thank God.

Found Clement and young George Darwin at Barton.*

* George Darwin was reading with him.

Strolled about the grounds with Fanny.

———

Monday, 6th July.

Busy unpacking, arranging and settling ourselves at home. A long talk on business with Scott. We inspected the Ferns and called on the Percy Smiths and the Suttons.

———

Tuesday, 7th July.

Excessively sultry.

Read the Odyssey, the first 99 lines of book 10. The passage about the land of the Laestrygonians, v. 82-86 is obscure, and has been very variously interpreted; v. 86—"the ways of night and day are near together," — would seem to admit of the interpretation, that Homer had heard some vague report of countries in which the night was so short, as to bring the evening and the morning, as it were, into contact. So, if I am not mistaken, Froude explains it.

Fanny went over to Mildenhall and returned to dinner.

══════

LETTER.

Barton,
July 7th, 1868.

My Dear Henry,

We returned home last Saturday, the 4th, and though I had expected to find the grass

E E

1868. parched and burnt up, the reality has even exceeded my expectations. The lawn and the park (except directly under the shade of some of the large trees) have the colour of a Cape landscape in the height of the dry season: the ground is as hard as a rock, and the grass so dry, that I wonder how the poor horses can get any nourishment from it. On the other hand, the trees are in their glory, their foliage remarkably fine; I think I have seldom seen it so rich and massy; and even the young lately-planted things look well.

Your Indian Horse-chesnut (the one in the arboretum) *has* flowered abundantly: but its flowers are almost entirely past, whereas last year it was in full blossom in the middle of July. The one near the stables is flowering well this year, and is not so far advanced as the others. The Ferns under the north side of the wall in the flower garden are surprisingly luxuriant, and those which you sent me from Wales last year are thriving, but in so unfavourable a season, I do not think it safe to plant them out. The long perseverance of this drought is certainly very extraordinary, and is becoming really alarming. If it continues much longer, I do not know what we shall do for food for our cattle: for no *roots* can be grown in such weather. We shall have to use up our winter store, and at last to kill the beasts to save them from dying of hunger and thirst. By *we*, I do not mean this parish in particular, for we are not at all worse off than our neighbours—perhaps rather better. We must hope that St. Swithin will bring us a good downpour.

I fully expected that our return home would be the 1868.
signal for heavy rains, as I had been often thinking
during the hot weather in London, how pleasant
it would be under the trees at Barton:—but the
rain still holds off.

I was very glad to return to this place, yet I
enjoyed our two months in London more than any
stay that I have made there for I do not know how
long. The pleasure of it was much increased by
our having those two most especially charming girls,
Sarah and Kate Hervey staying with us, first one
and then the other nearly the whole time. We had
a good deal of pleasant society, and not *too much* of
it,—at least, *I* took it very moderately, going
neither to balls nor *drums;* we saw several old
friends, and made several very pleasant new acquaintances.

I was very glad to renew my acquaintance with
Sir William Codrington, and to be introduced to
Lady Codrington. But above all, I was quite
fascinated by a lady whom we met first at Mrs.
Young's—Mrs. Walrond (Miss Grenfell) a niece of
Mrs. Kingsley: she is perfectly charming, and I
quite lost my heart to her. Her husband is
secretary to the Civil Service Commissioners, a
clever man and of pleasing manners.

I have not time now to dwell more upon our
London experiences. I was not entirely idle, for I
corrected the proofs of 300 pages of my Memoir of
our father, and also read three books of the Odyssey
in Greek. I was unwell the last few days we
were in town (as a great many other people were at

1868. the same time), and the bitter cold wind blowing in
my face as we travelled down on Saturday, gave
me an attack of rheumatic face-ache and headache,
but I am right again now.   Fanny is well, and I
need not say that she is busy.

Clement is here reading mathematics with George
Darwin (son of the famous Charles Darwin), who
was second wrangler.

A great deal of love to dear Cissy and the darling
children.   I look forward with very great pleasure
to the seeing you all here in September.

Ever your very affectionate brother,
CHARLES J. F. BUNBURY.

JOURNAL.

Wednesday, 8th July.

Mr. and Mrs. Percy Smith came to luncheon:
and I walked round the grounds with him.   Fanny
went to see Lady Arthur Hervey.

Thursday, 9th July.

As hot as ever,   Scott gave me good news about
the settlement of the Mildenhall shooting.

We travelled by post to Brandeston Hall, three
miles from Framlington, Mr. Charles Austin's.

Friday, 10th July.

The same hot and brilliant weather.   We all
went into Framlingham to see the Suffolk Agricul-

tural Association Show. Mr. Austin went to the 1868. public dinner at Framlingham. We strolled about with Mrs. Austin, her sister and children: saw the Church, &c.

————

We returned from a visit of a day-and-a-half to the Charles Austins at Brandeston Hall, about three miles south-west of Framlingham. They are agreeable. Mr. Austen is much softened by age and prosperity, though he still shows from time to time a good deal of the sarcastic and cynical humour for which he was so remarkable when I first knew him long ago, when I used to meet him in company with poor Charles Buller. He is a man certainly of powerful intellect, and of very extensive reading, and various information ; especially fond of classical studies. His wife is handsome and well informed, very pleasing and interesting ; sings delightfully.

Brandeston Hall was bought by Austin abont 21 years ago from the Revetts. It was an old house of Henry VI.'s time, but the walls (or part of them) are all that remain of the old house, for it was burnt down in 1848, and has therfore been entirely restored — indeed partly rebuilt — since that year. It is now a very handsome red brick house, in the Tudor style, with many gables, mullioned windows, &c. ; the interior fitted up to correspond ; and all as far as I can judge in good taste; but the mullioned windows appear to me, as in similar cases, more ornamental than convenient.

The little stream of the Deben winds through the

1868. park, and there are some very fine trees; especially a group of limes, near the house, and a superb elm, of the variety with smooth leaves and slender pliant twigs, of which we have many here. This is one of the largest and most beautiful trees of the kind which I have seen.

In the last few days, I have read a very clever, lively French book, called "Le Prince Caniche," by Laboullaye. Under the disguise of a fairy tale, it is a most keen and admirable satire on the prevailing faults of the French system of Government; the mania for centralization and for over-government; the incessant interferences; the passion for military glory; the distaste for tranquil prosperity. It contains some useful lessons for certain English political theorists, as well as for French statesmen.

Sunday, 12th July.

We went to morning Church and heard a good sermon by Mr. Percy Smith.

Monday, 13th July.

The hot, dry, brilliant weather continuing.

Read the Odyssey, book 10, verse 100-163. I think it has been already remarked, that there seems to be a want of consistency in the account of the Laestrygonians; they are described as giants, yet the comrades of Ulysses seemed not to have remarked this at first, nor to have seen anything to

alarm them till they arrived in the presence of the 1868.
King and Queen.*

Arthur Hervey and his two dear girls, Sarah and
Kate, came to dine with us, and walked about the
grounds before dinner. The Percy Smiths also
dined with us.

LETTER.

Barton, Bury St. Edmund's,
July 14th, 1868.

My Dear Lyell,

Thanks for your little note. I will order
Dr. Dawson's "Acadian Geology" (everybody in
this country will suppose it to be *Ar*cadian), and
shall like to see his latest conclusions as to the fossil
plants.

Cordaites is a genus of Unger, founded upon
Flabellaria borassifolia of Sternberg, and near to
(not very distinct from *Noeggerathia* of Sternberg
and of Brongniart). The latter, and Geinitz, also
considered it as nearly allied to Cycadeæ; but
Carruthers, I believe, doubts whether any of the
coal plants were of the Cycad family. Most of the
so-called *Poacites* and *Cyperites*, I supposed to be
leaves of Sigillaria.

Fanny is at this present moment, resting under
one of the trees on the lawn. I am reading your
43rd chapter with great care, and mean by-and-by

* This remark is made in Gibbon's " Extraits de mon Journal." (Miscellaneous Works, volume 5, page 417).

1868. to send you a letter heavily charged with doubts and difficulties.

I am much pleased with George Darwin.

With much love to dear Mary,

Ever yours affectionately,

CHARLES J. F. BUNBURY.

JOURNAL.

Thursday, 16th July.

Leonora and George Pertz and their dear little girls arrived.

Saturday, 18th July.

My dear wife's 54th birthday.   Joanna and Mrs. Byrne arrived.

Sunday, July 19th.

We went, a large party, to morning Church.

I corrected the last proof-sheet (the table of contents omitted) of my Memoir of my father. Before the copies are ready to distribute, it will be very nearly two years since I began to write it.

The wheat harvest is now beginning pretty generally around us here, and especially on the north side of Bury, where the soil is lighter than on this side. It is uncommonly early, and promises to be very good.

The continued heat and drought of the season are very extraordinary; I hardly remember anything like them at this time of year.   When we came home from London, a fortnight ago, the lawn and park here appeared parched and burnt up to a surprising degree ; as brown as a Cape landscape. Since then hardly a drop of rain has fallen, hardly a drop of rain In July, usually a rainy month. The lawn and pastures of course are more parched than ever ; they look absolutely whity-brown ; one is inclined to fear that the grass will die altogether. The farmers are in despair about their turnips and other root crops, and consequently alarmed about their stock. There are fears even of the supply of water for the gardens. On the other hand the foliage of the trees is remarkably fine ; as rich and luxuriant as I have ever seen it.

————

Monday, July 20th.

The wheat harvest begun on my farm, and generally about us.

————

LETTER.

Barton,
July 20th, 1868.

My Dear Edward,

I enclose a letter which I have received from a Mr. Darvel Gurteen, a manufacturer at Haverhill, on the subject of the representation of Suffolk. I have of course declined his proposal as far as. I am myself concerned—first, because I am Sheriff of the County ; secondly because on account of the state of my health and my deafness,

1868. I have for some years past, quite made up my mind
that I had better have nothing to do with public
life ; thirdly, because I do not believe that my
opinions when they come to be known would at all
harmonize with those of these "advanced Liberals."
I have written to this effect to Mr. Gurteen, at the
same time thanking him for his courteous expressions. I send you his letter, because I think it
possible that you may be inclined to feel your way
towards standing for West Suffolk, as you will see
that he is inclined to support a Candidate of our
family. He is not alone, for I have heard reports of
a similar movement at Stowmarket and that part
of the country.

It is for you to consider whether you think it
worth while to stir ; at the same time I cannot
promise you any active support. I am glad that
the "advanced Liberals" (or Radicals) are turning
their thoughts towards the old county families, and
the descendants of the old Whigs, and not to new
stump-orators ; and it would be well if this feeling
could be encouraged.

I should be glad enough too, to see Major Parker
turned out, but of this, I fear there is little chance ;
and I should not be glad to see Lord Augustus
turned out.

I told Mr. Gurteen in my answer that I should
send his letter to you.

I hope you are thoroughly enjoying that lovely
place, Abergwynant. I wonder whether it is as much
burnt up as we are here. The foliage of the trees
is remarkably rich and fine, but our lawn and park

are regularly whitey-brown. Give my love to
Henry and Cissy, and the dear children, and
believe me,

Ever your very affectionate brother,

CHARLES J. F. BUNBURY.

JOURNAL.

Wednesday, 22nd July.

Fanny and Mrs. Byrne went to dine at Hardwick.
Katharine and her husband and children arrived in
the evening.

LETTER.

Barton,
July 22nd, 1868.

My Dear Edward,

I have had a visit this morning from Mr.
Manning Prentice, a man in great business at
Stowmarket, a zealous Dissenter and a Radical;
his object was the same as that of the letter from
Mr. Gurteen, which I sent to you on Monday,
namely, to sound me about standing for West
Suffolk in the "Liberal" interest. I gave him to
understand that it was quite out of the question as far
as I was concerned, and I showed him a copy of my
answer to Mr. Gurteen; but I told him that I did
not know whether you might be disposed to come
forward, and I gave him your address in case he
should be disposed to write to you—which I think
he probably will.

1868.   I understand that Mr. Prentice is a very good
man, and of considerable influence among the
Dissenters.   His political opinions are evidently
quite different from mine; he professes indeed to be
a follower of Gladstone, but he hopes (as I fear)
that Gladstone will become progressively more and
more "Liberal," *i.e.* Radical.

But I could not make out that he had any good
groundwork for trustworthy assurances of support to
be offered to a candidate of his choice.   His information about the registered voters appeared to be very
vague ; indeed the registration of the new voters is
not completed, and he did not know even approxi-
mately, as far as I could make out, what the
addition to the constituency would amount to.   And
then though the Dissenters may be in preponderating
force at Stowmarket and Haverhill (Mr. Prentice
said these were the two centres of the movement),
this goes but little way towards making a strong
party in the county.

It is very doubtful, I think, whether the main body
of the farmers, great or small, would be much
influenced by an agitation proceeding from this
source.   If I were in your place, and wished to
be in parliament, I should be very cautious of
coming forward on the strength of the support of a
party in these small towns, before being assured of
a strong party in the purely agricultural population.

It will of course be for you to make enquiries and
judge for yourself.   Mr. Prentice told me it was
believed (in his party I suppose), that Major Parker
would not stand a contest, but would retire in case

of a contest. *If* this be true (but I doubt it), 1868.
it might make some difference as to the prospect ;
for he is the especial pet of the small farmers.

Ever your very affectionate brother,
CHARLES J. F. BUNBURY.

JOURNAL.

Thursday, 23rd July.

A change of weather. Very rough and cold wind.
Observed a young cuckoo. Went, a large party of
old and young, to the charity bazaar at Ickworth.
Saw all the Herveys, Mrs Abraham and Lady
Cullum.

Saturday, 25th July.

Emily Napier arrived, and Mr. Clarke of Trinity
College.

Sunday, 26th July.

We went to morning Church and heard a good
sermon from Mr. Percy Smith. Showed Emily
Napier the arboretum.

Monday, 27th July.

The Arthur Herveys came to afternoon tea, and
we lounged about the arboretum with them.

Tuesday, 28th July.

Most sultry.
Mr. Clark, Trinity College, tells me that near

1868. Vostitza, in the Morea, he saw a Plane tree that
measures 45 feet round the trunk. Nothing remains
of the stem but bark, yet the foliage is still fresh
and abundant. He tells me also, that some ancient
paintings of considerable interest, have been discovered in the ruins of a villa of the Empress Livia,
near Rome.—Near as I understand him, to the
place marked in the map as Saxa rubra, or Ad
Gallinas.

In particular, the representations of trees and
flowers, appeared to him more natural and truthful
than in any other ancient paintings which he has
seen. The Abrahams and Lady Cullum dined
with us.

Wednesday, 29th July.

A blessed shower of rain in the night. Fanny
and most of the party went to Mildenhall and
returned late.

Thursday, 30th July.

Again very hot.

Friday, 31st July.

Emily Napier went away. We went, a large
party, to Ickworth, and spent the afternoon very
agreeably with the Arthur Herveys.

Saturday, 1st August.

George Pertz went away early. George Darwin
in the afternoon. Read some more of Darwin on
"Variation." A long visit from Colonel and Mrs.
Ward.

Monday, 3rd August.

Still the same intensely hot and bright weather.

— — — —

Tuesday, 4th August.

Mrs. Byrne went away.

I received 25 copies of my Memoir of my father, and distributed some. It is a satisfaction to me to have thus brought to completion a work, which I began (almost exactly two years ago) with but little hope or expectation of ever finishing it, and about which my courage repeatedly flagged during its progress. I cannot estimate my own work at all fairly at present, for, having been so long engaged correcting it, I feel almost sick of the whole thing, and mean to lay it aside entirely for a good while. But I wish I may have succeeded in doing some kind of justice to *him*.

— — — —

Wednesday, August 5th.

Leonora and her children and Joanna went away in the morning, also Clement.

— — —

Thursday, August 6th.

The wheat harvest is now, I believe, completely finished in this parish; and yesterday, Mr. Cooper gave the harvest dinner to his labourers: the first time within the memory of man (as I understand) that this has happened so early.

We travelled post, taking Leonard with us, to Ipswich, left him there. Visited the Mills at Stutton: then to Woolverstone to the Berners.

1868. Mr. and Mrs. Charles Berners staying in the house. Lady Anstruther dined at Woolverstone—very pleasant.

_____

Friday, 7th August.

We drove into Ipswich, found the Judges were coming later. Went back to luncheon; drove into Ipswich a second time. Judge Keating arrived— Commission opened. The Austins, Sir Charles Rowley and his son and Captain Spencer at Woolverstone.

_____

Saturday, 8th August.

We drove early into Ipswich. Went to church with the Judges—their remarks on the service;* then to the Court. Trials of prisoners. We returned to Woolverstone in time for dinner.

_____

Sunday, 9th August.

We went to morning church. I walked with Mr. Berners through the Fernery.

_____

Monday, 10th August.

The Assizes—in court all day—great heat and bad air. Saw Mrs. Austin and Mr. and Mrs. Mills at luncheon at Fanny's lodging.

---

*The Judges were Sir Alexander Cockburn and Judge Keating. The service very high church; so when they got into the carriage to go to the Court, Cockburn told the Chaplain, Mr. Lott, that he would not go to the Church again to hear "such tom-foolery," and told him to find another Church for Sunday. Mr. Lott thought it better not to have any service for fear of giving unnecessary offence, so the Judges spent their Sunday at Felixstowe, and Mr Lott preached no sermon.

Again in Court all day from 9.30 to past 7—in horrible heat and bad air. The Chief Justice occupied the whole day in the trial of one case.

Met Charles and Mary and Leonard at Fanny's lodgings, and afterwards at the White Horse.

———

Wednesday, 12th August.

Released from attendance on the Judges. Took leave of our kind friends at Woolverstone and travelled home; arrived safe and sound and found all well, thank God. Though I had been occupied all day in the Courts at Ipswich, yet I had passed the evenings and nights under the roof of our kind friends, the Berners at Woolverstone.

Found at home a very pleasant letter from Cissy. It is a great satisfaction and comfort to me that Henry and Cecilia are much pleased with my Memoir. This day I began to read Lanfrey's "Histoire de Napoleon Premier."

Towards Ipswich I observed that the face of the country in general was quite as much parched as here; the deciduous trees fast shedding their leaves, and some, especially poplars and chesnuts, almost leafless.

———

Thursday, 13th August.

Weather a little showery.

Katharine with her two children arrived from Mildenhall.

There have been some refreshing rains in the last

1868. week, and the grass is already beginning to look greener.

Mr. Whelan, of the National Provincial Bank at Bury, has sent me a living specimen of a new British plant—Aster salignus—found (as he tells me) in Wicken Fen, Cambridgeshire, by Mr. Thomas Brown of Cambridge. There is no improbability of its being a native of England, as it inhabits the North of Germany, and especially the banks of the Elbe down to Holstein, but it is very strange that it should never have been found till now.

_____

Saturday, April 15th.

Pleasant letters from Arthur Hervey and Mr. Mills.*

We dined with Mr. and Lady Susan Milbank at Ashfield. Went early and saw their garden. Met there Mr. Tyrell, Mr. Thornhill the younger and Lord Thurlow.

_____

August 17th, Monday.

I sent Spottiswoode a cheque for the amount of their bill (£80 18s.) for printing my book. It has cost less than I expected.

_____

Tuesday, 18th August.

We went to stay with our good friends the Evans Lombes at Great Melton, near Wymondham, during the Norwich Meeting of the British Association. Mr. and Mrs. Webb of Newstead Abbey were staying in the house. Lady Bayning came to dinner.

* Probably on his Memoir of his Father.—F.J.B.

We went into Norwich after luncheon to the reception room, got our tickets. Went to see Mary and Katharine and Sedgwick. Dined with the Lombes at the Norfolk Club. We all went after it to the Drill Hall to hear Joseph Hooker's inaugural address, as President,—we got very good places, and heard it well. The address was excellent. The leading topics were, at the beginning, a notice in very good taste and feeling, of the first meeting of the Association which he had attended, and of the changes since then. Secondly, observations on the British Museum and on Provincial Museums. Thirdly, (and principally) a most excellent account of what Darwin has done for botany, leading to a notice of his work on "Variation," and a general defence of his theory. Lastly, observations on the supposed antagonism between science and religion, and the dislike or distrust shown by the clergy towards science. In this part I thought he was too lengthy and too defiant in his tone towards the clergy. But altogether it was a fine discourse. I was much struck with his notice of the *megalithic* monuments on the Khasia mountains in India. He said that the natives of those mountains (the Khasias), even to the present time go on erecting temples of enormous unknown stones, almost exactly like those which are called Druidical monuments in England and in Bretagne. These were first described by Colonel Yule, as long ago as 1844; in consequence of the representations of the British Association,

1868. systematic measures are now to be taken to obtain
and preserve all possible knowledge relating to these
people and their works, and it is an object well-
worthy of attention.

From my approbation of what Hooker said about
Darwinism, I must except the notice of *Pangenesis*
on which subject he appeared to me to be as
unintelligible as everyone else who has touched
upon it.*

The Mayor of Norwich spoke remarkably well in
seconding the vote of thanks to Hooker.

———

Thursday, 20th August.

A thoroughly wet day; the first for I do not know
how long. We went into Norwich to Section C;
heard a very long paper by Mr. Osmund Fisher on
the "Denudations of Norfolk." Charles Lyell
spoke on it very well, with great spirit and clearness.
Among other things, he mentioned that it was just
fifty years since he first made a geological
examination of the coasts of Norfolk and Suffolk,
in company with a Dr. Arnold, who had been with
Sir Stamford Raffles in the East, and that they
were much puzzled by the glacial deposits, as at
that time it had not occurred to anyone to speculate
on the action of ice. He mentioned also the
abundance of great icebergs which he himself had
seen floating in the Atlantic, about the latitude of
Norfolk, and reminded us of the great difference of

* Hooker's Presidential Address has been published, and I have read it
carefully, and certainly like it better in reading than I did in hearing.—
(F. J. B.)

climate between this country and the same latitudes 1868. in America.

Afterwards we went to the Institute of Prehistoric Archæology, to hear Lubbock's Address. Then we went to see Mary and Katharine, Mr. Symonds and Sir William Guise.

Again into Norwich to Section C. Pengelly read a detailed report on the exploration of Kent's Cavern, Torquay—very curious. Afterwards the same Mr. Pengelly spoke extremely well. A paper on the condition of the bones found in the said cavern, pointing out the four different conditions in which they are found: entire, crushed, broken across and split length-wise: contending that those which are broken across, had been so broken by hyænas, and that those which were split length-wise, had been so split by man, and giving an amusing account of the experiments he had made on the treatment of bones by living hyænas, and on the possibility of splitting them with implements of stone and wood. Boyd Dawkins and others afterwards spoke on the same subject.

Then we went to Section D, and heard a very lively and amusing speech from Frank Buckland, against the indiscriminate destruction of birds, and especially birds of prey.

We visited Sedgwick, and dined with the Lombes at the Norfolk Club. In the evening we went to the Drill Hall to hear Mr. Fergusson's lecture on

1868. Buddhism:—it was excellently well delivered, and
very interesting, but the subject was so new to me,
so little connected with any of my previous studies,
that I do not feel sure that I understood it
sufficiently, to retain clear and correct ideas of its
substance, and therefore I will not attempt to write
it down.

Saturday, 22nd August.

Furious wind and rain. Fanny's headache and
the bad weather prevented us going into Norwich.

Dr. Ewing who is at present acting as curate at
Melton, and who dined at the Hall this day, is a
very intelligent, well-informed, active-minded man,
a zealous ornithologist, and fond of natural history
in general. He formerly lived several years in
Tasmania, and ascended Mount Wellington in
company with Charles Darwin, and also with
Joseph Hooker. He had also resided some time at
the Cape, so that we had much to talk over
together.

Sunday, 23rd August.

Letter from Cissy: sad account of dear little
George. Severe disappointment about her visit
to us.

Monday, 24th August.

We went into Norwich to Section E. A long
paper by Mr. Whymper on Greenland. After it
Charles Lyell gave us an excellent discourse, or
lecture, on the same subject. He explained the

indications which remain to show that, in the 1868. Glacial Period, the cold of Greenland must have been greater, and the accumulation of snow and ice much greater even than at present. And he dwelt more particularly on the proofs of the more extraordinary change which must have taken place since the Miocene Age, when, as the fossil plants show, Greenland must have been a well-wooded country with a mild climate and a rather rich flora. He gave some account of these fossil plants, and of the researches of Heer, whom he praised with great justice.

Next, a paper on the Seychelles islands by Professor Perceval Wright—curious.

One remarkable botanical fact he mentioned, that he had found near the summit of one of the islands, a genuine species of *Nepenthes*, a genus never before found out of Asia. The Seychelles islands are the tops of a group of nearly submerged granitic mountains, each encircled by an annular coral reef, and having, as usual, calm water between the reef and the land. The lower parts of the island are luxuriantly fertile; all the finest tropical productions are successfully cultivated there, and many have become wild. The islands are nearly exempt from hurricanes, and the climate, though hot, is healthy. In fact, Dr. Wright described this group as an earthly paradise.

We went with the Lombes and Webbs to Mr. Harvey's Fête at Crown Point—great multitude of people: fireworks, home very late, very hungry and tired.

Tuesday, 25th August.

We parted from our kind friends at Melton, and
returned home: travelling with the Charles Lyells
and Webbs from Wymondham as far as Thetford,
where our carriage met us.

Both Mr. and Mrs. Webb are very pleasant. *He*
has travelled much, and has a good deal of information. Charles Lyell talked to me of a remarkable discovery made in Sweden, and communicated
to the Association at Norwich by Professor Torrell,
the discovery of fossil vegetable remains in the
Lower Cambrian rocks (corresponding to the Long-
mynd group in England). He says they are
believed to be remains of *Monocotyledonous* plants,
and have been compared to Acorus Calamus, so
that their discovery makes a remarkable change in
our notions as to the earliest forms of vegetation.

All well at home, thank God.

———

Wednesday, 26th August.

Arranged a beautiful new writing table which
Fanny has given me.

A visit from dear Sarah Hervey: Katharine and
Leonard, Herbert Bunbury, Sir William Guise,
Mr. and Mrs. Bentham and Mr. Symonds arrived
from Norwich.

———

Thursday, 27th August.

Showed some of my dried plants and fossil plants
to Mr. Bentham, Sir William Guisc and Mr.
Symonds—much good botanical talk.

We all went in two carriages into Bury and saw the 1868 Churches. Lady Cullum dined with us.

Friday, 28th August.

Mr. Pakenham Edgworth arrived. I walked through the garden and arboretum with Mr. and Mrs. Bentham, and Mr. Edgworth. Mrs. Abraham and the Percy Smiths dined with us.

======

LETTER.

Barton,
August 30th, 1868.

My dear Henry,

I am indeed most sincerely grieved to hear such a sad account of dear little George as your last letter gives me, and I feel deeply for the constant and painful anxiety which his state must cause to you and dear Cissy. I can only pray that it may please God soon to relieve your sorrow and distress. If you should think it advisable to have the opinion of Dr. West or any other doctor from London, do not let the expense be any impediment, for I should be most happy to take it upon myself. I must not, when your anxiety is so severe, allow myself to dwell upon my own disappointment: but it is a very great disappointment to lose the hope of seeing you or any of you here this autumn. There are so many things I had hoped to show you, and to have your opinion of. But we must trust to see you all under happier auspices next year.

1868.   Your opinion of my book is very gratifying to me.
I am well aware how far I have fallen short of my
own idea and of full justice to the subject, but I
have done my best.

We spent a very pleasant week in Norfolk, with
the Evans Lombes, going almost every day into
Norwich to attend the British Association, where
we met many friends and acquaintances. Since we
came back we have had a detachment of the
learned men from Norwich visiting us here:—Mr.
Bentham (the President of the Linnean Society),
Sir William Guise, Mr. Symonds, Captain Brine
and Mr. Edgworth* who is here still: altogether
very pleasant. Katharine and her children were
with us till yesterday, when they set out for London
on their way to Tenby.

We met at the Lombes a clergyman, who was
formerly for some time in Tasmania, and remembered seeing you at Eagle Hawk Neck :
his name is Ewing: he is a very intelligent
man, a zealous naturalist, and ornithologist in
particular.

Mr. Edgworth who is a distinguished Indian
botanist, was quite delighted at the sight of your
Indian Horse-chesnut ; he did not know it had been
introduced into Europe, and never expected to see
it again: he quite agrees with you as to its extraordinary beauty. He measured one in the
Himalayas which was above 40 feet round the
stem.

* Mr. Pakenham Edgworth.

I am in pretty good condition I am thankful to 1868.
say. With very much love to Cissy and all your
children.

> Believe me,
> Your very affectionate brother,
> CHARLES J. F. BUNBURY.

=====

JOURNAL.

Monday, 31st August.

Mr. Edgworth went away early. We have had a
pleasant party of scientific men staying with us on
their way from the meeting at Norwich.—The
Benthams, Sir William Guise, Mr. Symonds of
Pendock, Captain Brine and Mr. Pakenham
Edgworth, besides Katharine and her son Leonard;
Mr. Bentham and Mr. Egerton are particularly
pleasant. We had much good talk.

Several of Mr. Bentham's botanical remarks I
have noted down in my scientific note book. He
said that he had observed, both about Norwich and
here, that some of our common plants appear to be
such as are wanting or scarce in Herefordshire; he
instanced Salvia Verbenaca and Lycopus Europæus.
Mr. Bentham said that the genus Benthamia is
hardly tenable; he thinks it must merge in Cornus,
as the only difference is in the fruit, and there is
a species of Cornus (capitata) in which the fruits
are almost confluent. I objected that on this
principle he would break down half the old genera,
and asked him whether he did not think that if we

1868. were perfectly acquainted with all the plants that have ever existed, we should find almost all the genera and groups of whatever grade, shading off and passing into one another. He admitted this, but thought that the distinction between a genus in which the fruits are confluent, and another genus in one species of which the fruits are *nearly* confluent, was too finely drawn. Perhaps he is right, as there is no difference of habit, and the Benthamia is a solitary species.

Mr. Bentham agreed with me, that the rule of priority of names, is now pushed to a very inconvenient and troublesome excess. Mr. Bentham told us that when he travelled in the Pyrenees with Mr. Arnott, many years ago, they collected and preserved 30,000 specimens of plants in three months. They began operations by buying *twenty reams* of soft French paper.

Bentham remarked that Agassiz has told us near the beginning of his "Journey in Brazil," that one of his main objects in that expedition was to collect facts by which he might refute the "transmutation" (or Development) theory, but he nowhere in the book says a word by which one can judge whether he thought himself successful in this object or not.

Our visitors were much pleased with the arboretum here, and particularly with the Catalpa and the Pinus excelsa, which both Mr. Bentham and Mr. Edgworth pronounced to be among the finest trees of their respective kinds that they had seen in England. Mr. Edgworth was much pleased at seeing the Æsculus Indica; he had not seen it 1868. since he left India, and was not aware that it had been introduced into Europe; indeed he thought there would be much difficulty in introducing it, on account of the oily nature of the seeds. He says it is one of the most beautiful trees he knows, and grows to a vast size; he measured one in the Himalayas which was upwards of 40 feet round the stem. It ascends, he says, nearly to the snow.

He has measured Deodaras 46 feet round. The Cupressus torulosa also, he says, grows to an immense tree, and of a very grand character. Mr. Bentham said, one of the largest trees he ever saw was a Plane, in one of the courts of the Seraglio at Constantinople; I cannot remember exactly what he said was the circumference of the stem, but it was between 40 and 50 feet, therefore somewhere about the same size as the Plane at Vostitza, of which Mr. Clark told me. (See July 28). The famous Plane tree at Buyukdéré is much larger, but in that case it is very doubtful whether it is a single tree (Mr. Bentham).

Read to-day part of Adair's pamphlet on Ireland. Clement arrived.

———

Wednesday, 2nd September.

Read a paper by the Duke of Argyll in the *Quarterly Journal* of the Geographical Society on the Physical Geography of Argyllshire, in connexion with its geological structure; in fact, in opposition to Geikie's extreme theories of Erosion or Denudation. It appears to me that (as so often happens in cases

1868. of this kind, the combatants on both sides push
their theories much too far, and that the truth lies
between them ; in fact that both theories are par-
tially, and neither wholly true. The Duke, I think,
does not rightly estimate the effects of rain and
rivers ; Geikie and his party greatly under-rate the
effects of internal movements and igneus action.
In any volcanic country, such as Teneriffe, one
cannot fail to see how important has been the share
of each.

Read also part of the *Edinburgh Review* on Prince
Henry the Navigator, a very well written and
interesting article.

The Adairs, the Barnardistons, Lady Rayleigh
and her daughter and sons, and Catty Napier
arrived.

_____

Thursday, 3rd September.

Beautiful weather. All the party except myself
went to Ickworth.

George Napier arrived. The Arthur Herveys
dined with us.

_____

Friday, 4th September.

Splendid weather. We went (a party of sixteen),
to luncheon at Ickworth (the great house), and
to the shooting for prizes,—came home very late
and very tired.

_____

Saturday, 5th September.

Very hot and bright. I signed the lease of the

Mildenhall shooting to six of the principal tenants 1868. for £300 a year.

LETTER.

Barton,
September 5th, 68.

My dear Edward,

Mr. Manning Prentice and Mr. Daniel Gurteen are persevering men. They came here yesterday and paid me a long visit, wishing to induce me to persuade you to stand for West Suffolk. They have either in person or by agents, been making a detailed examination of all the electoral districts and almost all the parishes of West Suffolk, classifying the voters now on the register (new as well as old), according to their supposed political leanings, as Liberals, Tories, or Doubtful; and they think that the result of this reckoning justifies them in wishing to contest the Division, and gives them reasonable grounds for hoping for success. They are particularly anxious to try their strength in *this* general election, the first under the new Reform Bill, and that the new voters should have at once an opportunity of recording their opinions, and they think that so favourable an opportunity is not likely to occur again. They wish particularly to bring forward as their candidate a member of one of the old Whig families, and *especially* (they say) a Bunbury: they declare that a Bunbury would have many more votes than anyone else.

464 MEMORIALS OF SIR CHARLES J. F. BUNBURY.

1868.   Upon all this there are, as it strikes me, several things to be observed. On the one hand, I cannot help being pleased by the good feeling they show towards our family, and I approve very much of their wish to bring forward a member of one of the old county families, and not a stranger. They seem much in earnest, and it may be apprehended that, if they fail in finding a gentleman candidate to their mind, they may bring forward a farmer, as the Tories did in Norfolk. On the other hand, it appears to me certain that the contest would be (to take the most favourable view) a very doubtful one. Mr. Prentice and Mr. Gurteen are evidently very eager and very sanguine men, and I should think them very likely to deceive themselves and to make serious mistakes as to the strength of their party. I should strongly suspect that their calculations would require to be received with great caution. You remember how our Father was misled in 1837. They did not mention to me the name of any one man of note or importance in the county whom they could reckon as co-operating with them. And then the question of *money* is a very serious one. I gathered from the conversation that the expense of contesting the Western Division would probably not be less than £2000: and I suspect it might be a good deal more, as it seems that elections in general this year, are likely to run to a great expense. This might probably be inconvenient to you: and though I should like to see you representing West Suffolk, I should much hesitate to spend so large a sum on (what has always appeared to me a peculiarly unsatisfactory investment of capital) a 1868. contested election, not to mention that it would hardly be decent for the Sheriff of the county to be supplying funds to one of the candidates.

We spent a very pleasant week with the Evans Lombes, during the British Association Meeting at Norwich, and met at their house, Mr. and Mrs. Webb of Newsted Abbey, pleasant people.

After returning home, we had a detachment of the learned from Norwich, staying with us a few days; the Benthams, Mr. Pakenham Edgworth, Sir William Guise, Mr. Symonds and a Captain Brine; all pleasant, especially Mr. Bentham and Mr. Edgworth. Since then we have again had a house full of company, for the rifle contest at Ickworth, but to-day we have none remaining but the Adairs, Catty Napier and George.

I know Fanny wrote to you a few days ago, and I conclude she has told you more particulars about our Norwich experiences.

Poor Cissy and Henry are in sad anxiety about dear little George, and though I trust he will get over this attack, I am much afraid that he will for a long time be a cause of constant anxiety to them. I hope you are enjoying your rambles in the Pyrenees; if you have such weather as we have here it must be delightful.

Ever your very affectionate brother,
CHARLES J. F. BUNBURY.

JOURNAL.

1868.   Excessively hot.

Read Flower's paper in the *Quarterly Journal* of the Geological Society on "Thylacoleo," an extinct quadruped from the caves of Australia. The writer contraverts Owen's conclusion, that this animal was a carnivorous and indeed an especially predaceous and voracious one.   I am afraid the result is, rather to shake our faith in the determinations of the habits and affinities of animals from their teeth.

We had a large party at dinner—the Abrahams, Milbanks, Lady Cullum, the Wards and the Percy Smiths.

LETTER.

Barton,
September 9th, 1868.

My Dear Katharine,

I thank you very much for your letter from Tenby.   We had already heard from Mary, of the alarming illness of your brother-in-law,* and truly sorry I was to hear of it.   I feel very much for Charles Lyell and your husband, and still more for the sisters, above all for her who has been living with him, and to whom his loss would be the breaking up of a home.   I hope he may yet be

* Mr. Thomas Lyell.

spared, but there seems to be cause for great 1868. apprehension.

I am very sorry too for the interruption to your pleasant time at Tenby. I do hope Charles and Mary will not attempt a hurried journey to Scotland, I am sure they are not strong enough to bear it without injury.

I read with much interest your account of Tenby, and should like to be with you in botanizing excursions there. I have often seen the Soap-wort growing apparently wild, for instance in hedges in two or three places between this and Ipswich ; but always with double flowers, which is against its being a true native. It runs wild (or becomes *feral* as Darwin would say), so readily, and was so general in old gardens, that it is difficult to say where it is native ; but Hewitt Watson admits it to be probably so *on the coasts* of Devon and Cornwall, and perhaps of some other English counties.

Talking of Darwin, I have lately seen a curious collection of many kinds of Pigeons, belonging to Mr. Sutton's sons, at " The Cottage " here ; there are several of the varieties about which I have been reading in Darwin — Pouters, Fantails, Carriers, Tumblers, and some others, very curious and pretty.

Since the beginning of this month, we have again had the house full of company, but not of a philosophical cast (with the exception of John Strutt).*

This morning, the Adairs, who are very pleasant

+ Afterwards Lord Raleigh.

1868. have gone away, and we have nobody with us just now but dear Catty Napier, (a special favourite of mine), George Napier and Clement. Fanny and I are much as usual as to bodily health. The accounts from Abergwynant are rather better ; they write that dear little George is gaining ground slowly, but perceptibly ; but I am much afraid we have not much chance of seeing any of them here this year.

Are you free from wasps at Tenby ? here we are pestered by them to an extraordinary degree, and I think it quite a wonder that we have none of us been stung by them.

Our trees are already beginging to look quite autumnal. The Belladonna Lilies are flowering this year in remarkable abundance and beauty.

Pray give my love to all your party, and believe me,

Your very affectionate brother,
CHARLES J. F. BUNBURY.

JOURNAL.

Thursday, 10th September.

The Frederick Greys arrived, also Lady Anstruther. We walked about the grounds with them, and drank tea out-of-doors. Afterwards Captain Spencer and Mr. Boileau arrived. The Percy Smiths dined with us.

Friday, 11th September.

Very beautiful autumnal weather. I gave my book to Lady Anstruther, and walked about the arboretum with her. Went into Bury with the 1868. ladies. The Arthur Herveys and Lord Charles Hervey dined here.

Lady Anstruther, Sarah and Kate Hervey and dear Catty went away. The Benthams arrived.

A very cold N.E. wind. We went (a large party) to morning Church. Heard a good sermon from Mr. Percy Smith.

Lord Charles Hervey came to luncheon with us, and we showed him the ferns and the arboretum— he was very agreeable. Captain Spencer went away. Mr. and Mrs. Andrews arrived. The Hortons dined with us.

The Benthams went away. Arthur Hervey and Sarah, Lady Cullum and the Wilsons dined with us.

Mrs. Ferguson Davie and Lord Charles Hervey arrived. We showed them the fern-houses, arboretum and stables. Mr. and Mrs. Augustus Tollemache dined with us.

LETTER.

My Dear Edward,

1868.   I thank you much for your letter of the 10th, which has gratified me much by its tone of affectionate confidence. I quite understand your feeling of regret that this opportunity of contesting the county should be lost, though I cannot share in Shafto Adair's sanguine views as to the prospect: neither indeed can I much sympathize with his party eagerness. I assure you that I would advance the requisite money with quite as much readiness for you as for myself, if it were an object of primary importance in your eyes: if, in short, you intended to make parliamentary life your *career*: but this, as I conclude from your letter, is no longer the case; and unless for such a primary object of your life, I think the sacrifice would be too great. For you are quite right in supposing that it would be necessary to be prepared, not for one contest only, but for many. What the liberals want (as Mr. Prentice acknowledged) is a candidate who, if beaten in the first election, will try again and again and spare neither labour nor money,—one, in short, who will fight the battle for West Suffolk as Adair is fighting it for East. And, of course, this implies an indefinite expenditure.—To the fact that you are not an actual resident landowner, I do not attach so much importance as you do. You

belong to the country and are well-known in it, and 1868.
I do not think it would be easy to represent you
as a stranger. But I do not think that there would
be little use in you or any one else contesting
the Division unless you were secure of the hearty
support of some of the leading country gentlemen,
and I do not feel at all sure that any of them *would*
support a decided " Liberal." Maitland Wilson
has declared against the disestablishment of the
Irish Church, and I suspect that the Tyrells and
even the Rowleys are very lukewarm Whigs. As
for the farmers, (with the exception of a few like
Frederick Paine), I apprehend that the political
aspirations of the generality go little beyond the
repeal of the Malt Tax. As Fanny says, I think it
would be rash to come forward on the invitation
only of Mr. Prentice and Mr. Gurteen and their
respective cliques in Sudbury and Haverhill,
without a regular requisition including a good
number of good names of country gentlemen and
farmers as well as tradesmen. On the whole,
though I am half sorry that you should lose this
opportunity, I believe that you are quite right in
your decision. But do you not think that, if you
were in England just now, you would be in a better
situation for taking advantage of any opening
which might occur? For myself, independently of
my health and my distaste to a parliamentary life,
I feel at present much disinclined to take any active
part in politics, because I do not like any of the
political parties—and I could not exactly expect to
form one for myself. I feel no confidence in the

1868. present Ministry; I look upon Disraeli as a dishonest politician: yet I should have no satisfaction in helping to set up a Ministry which might include John Bright and John Mills, or at any rate would be completely under their influence.

I wonder whether the Pyrenean country is as much parched as we are here. When we first returned from Norwich in the latter part of last month, our grass had much revived and the lawn and park were qnite green, but the renewed drought which has been quite as severe as the first has again withered everything up. Lately we have had some very cold weather, but without a drop of rain, and the effect has been to bring on a premature autumn: the Beeches and even some of the Oaks have turned quite brown, and the leaves are falling as fast as if it were the end of October.

We have had a very pleasant party in the house: the Frederick Greys, Benthams, Catty Napier, George, Susan Horner and now Mr. and Mrs. Andrews and their daughter Mrs. Fergusson Davie. To-morrow our party breaks up. I hope you will enjoy the remainder of your tour,

Ever your very affectionate brother,
CHARLES J. F. BUNBURY.

JOURNAL.

Thursday, 17th September.

Our pleasant party broke up.—All our guests went away except Susan. We dined at Hardwicke —a large party.

The party staying with us, between the 10th and 17th was very pleasant—Sir Frederick and Lady Grey, Lady Anstruther (daughter of my father's old friend, Sir Henry Torrens, and a very agreeable and interesting person), my dear and) charming cousin Caroline Napier, Captain Spencer for part of the time, the Benthams, Mr. and Mrs. Andrews and their daughter, Mrs. Ferguson Davie, Lord Charles Hervey, who is very accomplished, and has much of the winning manners of his brother Arthur, and our constant favourites, Sarah and Kate Hervey. It was altogether a very agreeable time.

————

Friday, 18th September.

On the 10th of September, I examined the capsules of Hibiscus Syriacus and noted down my observations on them. It is the first time that I have observed this shrub to bear fruit here.

On this day (September 18th) I examined the flowers of a new and remarkable species of Begonia (Boliviensis) which has flowered in our hot-house, and wrote notes on it.

Lanfrey's remarks on Moreau's campaign of 1800 on the Danube, led me to read with care—Colonel Hamley's analysis of that campaign in his " Operations of War." It is extremely clear. He is not severe upon Moreau, but does not estimate him as highly as Lanfrey does. The military criticisms of the latter, do not appear to be of much value.

My Barton rent audit very satisfactory.

A little rain at last.

————

Saturday, 19th September.

A really wet day—very welcome. Lady Napier arrived.

———

September. 20th, Sunday.

We went to morning Church.

We began to make a list of the trees in the arboretum for the purpose of labelling them.

———

Thursday, 24th September.

Examined the structure of Cobœa scandens, and, made a note of it.

Read the Odyssey, book 11, v. 404-461. Agamemnon, or rather his shade, relates to Ulysses, the story of his murder. It is rather different, I think, from what is told by later writers, for nothing is said of Clytæmnestra killing him with her own hand in his bath ; but he speaks of himself and his comrades being murdered by Ægisthus, while at table.

Froude thinks (and I believe he is right) that the purpose of the poet in referring so often in the Odyssey, to the crimes of Clytæmnestra, is to bring out in stronger relief, the virtue of Penelope ; that Clytæmnestra was meant to be a foil to Penelope.

I walked round my farm.

———

Friday, 25th September.

A thorough good wet day.

Read the Odyssey, book 11, v. 462-540. Ulysses' conversation with the shade of Achilles. From beginning to end, nothing can be more melancholy, more gloomy, more depressing than the picture that Homer draws of the state of the departed, even of the greatest heroes. Even the Scandinavian idea of heaven, with its perpetual feasting and its bowls of mead, is cheerful in comparison with the dim vague, gloomy shadow which seems to hang over the dead in Homer's imagination. It rather enhances the merit and the valour of those old Greeks if they had nothing to look to after death but such a gloomy prospect as this.

———

Saturday, 26th September.

A beautiful day. News of the death of Dean Milman.

We labelled twenty trees in the arboretum. Fanny and Susan went over to Cockfield to see Mr. Babington.

———

Sunday, 27th September

Blowy and showery. We went on with our list of trees in the arboretum.

I read Kingsley's sermon on " The wages of sin."

The newspapers contain many particulars of the dreadful earthquakes which last month ravaged Peru, from the Equator to the Southern Tropic. It seems to have been one of the most extensively destructive series of shocks on record. The *earthquake wave* (the retreat and violent return of the sea) seems to have been especially destructive.

———

Stormy and wet.   Scott gave me a very satisfactory account of the Mildenhall rent audit.

———

LETTER.

Barton,
September 28th, 68.

My dear Katharine,

Many thanks for your letter. I was very glad to hear of your excursions, and how you have been spending your time at Tenby: and I hope it will have been on the whole a pleasant time for you, as well as profitable to your health; though the sad state of your poor brother-in-law must, I fear, have much damped your enjoyment.   The last account we had of him from his sister Marianne was very melancholy.   It must be a most painful time for all his sisters.

I was sorry and shocked to hear of Mrs. Yorke's sudden death—though I knew her but very slightly, because I knew that Mary had a regard for her, and that it would be an additional shock to poor Lady Head.*   Dean Milman's death too, will be a cause of sorrow to Charles and Mary, and all must feel a regret that so able and distinguished a man has been taken away, though in the fulness of years and honours.

Dear Susan left us this morning.   I am very sorry she would not stay longer: she was very pleasant and I think her looking very well.   For some time

* Mrs. Yorke, the wife of Colonel Yorke, who was Lady Head's brother.

past we have had no company but her and that 1868.
most sweet and amiable person Lady Napier, who
will remain with us till next Tuesday, the 6th. On
that day, or rather on the Monday, begins our
next *batch* of gaieties, and our most especially gay
gaieties, including the Bury ball and our own dance.

Fanny, as usual at this season, is hard at work
beating up for recruits for both.

This evening, (29th), we are going to the Bury
Athenæum, to hear Arthur Hervey's lecture, which
opens the course for this season. Our last party
was a very agreeable one: the Frederick Greys,
Lady Anstruther, the Benthams, Andrewses, Lord
Charles Hervey, &c.

I am afraid I have not a great deal to tell you,
though I do not think I am quite idle. I am
reading Heer's "Flora Fossilis Arctica," (in
German), and going on by degrees with Darwin on
"Variation," also with the Odyssey, and have
begun to read Kinglake's 3rd volume ("Invasion of
the Crimea,") which seems, as far as I have yet
gone, to be written with excessive diffuseness.

I shall be very glad to give you any further help
I can in your book of Ferns. We are having the
trees in the arboretum labelled with zinc labels, on
which the names are written by Dudley, and
beautifully he writes them.

I hope you have kept a list of the plants you
observed about Tenby: it is worth while whenever
one goes to a distant part of the country to note
what are the prevailing plants, for it often happens
that some of these are different in different districts.

1868. Mr. Bentham said that some of the common plants here and about Norwich are such as he was not accustomed to see in Herefordshire. So, I observed that the Hartstongue is the commonest of Ferns on the south side of the Isle of Wight and in the south of Devonshire, whereas I never saw a plant of it about Abergwynant. In this, as in many other things, I have grown wise too late: for I did not find out the use of local observation of common plants till my travelling days were over.

What dreadful earthquakes in Peru! the most *extensively* destructive that I remember to have read of.

Much love to your children,

Ever your very affectionate brother,

CHARLES J. F. BUNBURY.

JOURNAL.

Tuesday, 29th September.

We went with Lady Napier to the Bury Athenæum, and heard a beautiful and admirable lecture from Arthur Hervey on Napoleon. He had in previous years given two other lectures on the same subject. This is the concluding one—embracing the latter part of Napoleon's career, especially the Russian expedition; and it was perhaps the very best I ever heard from him, though not containing so much that was new to me as the one on Charlemagne.

Extremely wet. Went to Bury. Committee meeting of Convalescent Home (Felixstowe). Mr. Keene, Arthur Hervey, Maitland Wilson and I. We settled matters comfortably.

I called on Mr. Borton and discussed another piece of business with him. Read two biographical essays by Charles Kingsley, in *Good Words*, on " Rondelet" and " Vesalius" both very interesting, especially the one on " Rondelet."

Saturday, 3rd October.

Read the Odyssey, book 11, v. 540-640, the end of the book. It is very difficult to understand the ideas which the Greeks of Homer's time had of the state of the departed. Hercules we are told is himself in Olympus, among the immortal gods ; but his "eidōlon"—his image or spectre—appears to Ulysses among the other heroes evoked from Hades, and converses with him.

Mr. and Mrs. Augustus Tollemache came to luncheon, and to see the pictures.

Monday, 5th October.

Dear Minnie and Sarah arrived—also Mr. and Mrs. Mills, the two young ladies Legge, Captain Spencer and the Thornhills.

Tuesday, 6th October.

Dear Lady Napier went away. Lord John Hervey and others arrived. Our dance very

1868. successful.   Charming girls : everybody looked
pleased.   I danced Sir Roger de Coverley with Mrs.
Abraham.

Wednesday, 7th October.

A very beautiful day.   I had a very pleasant walk
with the dear Hervey girls and Mr. and Mrs.
Thornhill.

Thursday, 8th October.

Another beautiful day.   In the morning we went
with Mr. and Mrs. Mills and Mr. Thornhill to the
school, stables and farm.   In the afternoon I had a
delightful walk with the young ladies.   Fanny and
most of the party went to the Bury ball.

Friday, 9th October.

The young ladies and some of the men were
photographed.   I did some business.   Captain
Spencer, Lord George Pratt and Mr. Gurdon went
away.   Mr. and Lady Mary Powys arrived.   The
Arthur Herveys dined with us—merry games in the
evening.

Saturday, 10th October.

We were all photographed in a group.   Our
charming party broke up.   First Lord John and
John Hervey went away,—then the Thornhills,
lastly, our dear Sarah and Kate Hervey.   The
Wilsons and Lady Cullum dined with us.

Poor William Napier laid up with illness. No studies at all this last week. A week of gaiety and pleasant dissipation. Our house full of company including our dear Minnie, the Millses, Thornhills, Captain Spencer, and many others ; and a charming bevy of girls. The incomparable Sarah Hervey and her sister Kate, Sarah Napier, Lady Octavia and Lady Wilhelmina Legge, and two Miss Thornhills.

———

Monday, 12th October.

A most beautiful day. We planted trees in the Vicarage Grove for Mr. and Mrs. Mills, and for Minnie. I walked with the latter and Lady Mary Powys. The Millses and the Lady Legges went away. We and Mr. Powys dined at Ickworth with the Arthur Herveys. I sat next Sarah.

Finished Lanfrey's "Histoire de Napoleon I." I have been a long time about it, having been very idle and unsettled for the last two months. The two volumes hitherto published come down to the rupture of the Peace of Amiens, in the beginning of 1803. Lanfrey treats the military campaigns in a very brief and concise manner—in fact in mere summaries, and I do not think his criticisms on this subject are of much value. His special object is less to narrate than to analyse the career of Napoleon in his *moral* aspect, to supply the ground of a moral judgement upon him.

This he certainly does with a most stern and unsparing rigour ; without passion, but perhaps not with perfect impartiality, for he always puts every

1868. action of his subject in the most unfavourable light, and puts the worst interpretation upon it.

It cannot be denied that his view is somewhat one-sided; but even so it is a wholesome corrective to the flagrant partiality shown by the generality of French writers on the subject.

So many have laboured to deify Napoleon, that Lanfrey may well be pardoned for painting him perhaps a shade blacker than necessary. Be it observed too, that all the facts on which he bases his censures are derived from Napoleon's own correspondence.

The work gives certainly a very instructive and impressive picture of the intense egotism and selfishness of the great conqueror; of the wonderful persistency and force of purpose with which even from the beginning of his career, he laboured for his own glory and his own aggrandisement; making everything subordinate and subservient to these objects; disregarding all moral principles, setting aside all regard to the feelings and interests of others, if they stood in the way of his ascent to greatness.

Lanfrey endeavours to show, and I think he does show, that this intense spirit of self-seeking was not merely developed in Napoleon by his wonderfnl successes, but was characteristic of him from the beginning of his career.

Lanfrey has evidently strong Republican tendencies, and is disposed to judge more favourably of the first French revolution, and of the men of that revolution, than I should do. He always shows an in inclination to exalt the Republican Generals, such 1868. as Hoche, Kleber, and especially Moreau, in opposition to Napoleon, and to the Generals who were trained by him. Yet this Republican tendency does not prevent him from doing justice to the English, and especially to Pitt, whom he judges very fairly.

He makes it very clear too, that the rupture of the Peace of Amiens must really be laid to the charge not of the British Government, but of the first Consul.

Tuesday. October 13th.

We attended the Church Conference which met at Bury on the invitation of the Bishop of Ely.* It was very well attended, and the speaking was on the whole better than I expected. I was very much pleased with the address of the Bishop; it was most reasonable, temperate and mild; full of Christian charity and good sense; a true example of "*mites sapientia.*" Next to this I was most pleased with good papers and speeches from Arthur Hervey, Mr. Chapman of Bury, and our Barton Vicar, Mr. Percy Smith.

As might be expected, there was a considerable display of very different and conflicting opinions on the subjects discussed, and the speakers did not by any means all imitate the mildness and moderation of the Bishop.

Mr. and Lady Mary Powys left us.

---

* Harold Brown, afterwards Bishop of Winchester.

Wednesday, 14th October.

A long talk on business with Scott.

Frederick and Augusta Freeman arrived.

———

Thursday, 15th October.

Mr. and Mrs. Fergusson Davie arrived. I took a walk with them, Augusta and Sarah. The Percy Smiths dined with us.

———

Friday, 16th October.

William Napier convalescent. Edward arrived. Mrs. Hardcastle dined with us.

———

Saturday, 17th October.

We labelled thirty of the trees in the arboretum. The Fergusson Davies and Mr. Lott went away.

Edward issued his address to the electors of Bury.

———

Monday, 19th October.

Edward began his canvass at Bury. Lady Cullum dined with us.

———

Tuesday, 20th October.

Read the Odyssey, book 12, verse 73-141. So monstrous an idea as the description of Sylla in this passage of the Odyssey, seems rather Oriental than Greek.

Edward, who has just come from the Pyrenees, tells me that the *Rhune* mountain on the Bidassoa (so famous in the campaign of 1813-14) is very nearly of the same height as Cader Idris, just 1868. under 3000 feet. The Quatre Couronnes another mountain in the same group, and also a scene of fighting in the same campaign, is, he says, a mountain of a very fine and bold character, of something near the same height as La Rhune.

Wednesday, 21st October.

William Napier left us.

LETTER.

Barton,
October 23rd, 68.

My Dear Henry,

This week is a little lull in our autumnal course of dissipation, and accordingly, both Fanny and I have taken the opportunity to catch cold. I really have but little to tell you, for I know that Fanny has written to Cissy all about our gaieties— about our delightful society during the ball week, and especially about William's illness, and how he was consoled by bevies of nymphs brought around his sick-bed by some mysterious attraction.*

Edward is very busy canvassing. Lady Cullum says he ought to be regaled by *canvass*-backed ducks. All I can say is, that I am very glad he likes it:—I should not. As he has not put forward any very radical doctrines, I can heartily wish him

---

* He was sleeping in what is now called the smoking room, which looks into the pleasure ground and croquet ground.—(F. J. B.)

1868. success, but I fancy he will have a very close and stiff fight.

I do not think I ever saw the autumnal colouring of the leaves more beautiful than it has been this season, and indeed, still is. Now, indeed, a good many of the trees have nearly lost their leaves, but the Liquidambers are still most gorgeous with their crimson leaves, the Beeches are only now beginning to change colour, and the oriental Planes and some others are still quite green. The black Walnuts have a very odd look, with their leafless branches quite loaded with great green balls. The Catalpa, too, bears an extraordinary crop of pods, which hang down like long green candles. The berry-eating birds will have fine feasting this year, among the Hawthorns and Hollies.

Does Pinus (or Picea) *nobilis* thrive with you? I have planted two or three, but they have all died: yet Veitch says it is hardy.

I have just finished reading the third volume of Kinglake's "Invasion of the Crimea."

There is an article by Colonel Hambley (William tells me) in the new number of the *Edinburgh*, which is likely to be worth reading, as it may be expected to contain good military criticisms.

I hope dear little George is going on quite well. Give my love to all of them and especially to dear Emmy* whom I am very sorry not to have been able to see and talk with this year. I have no doubt, from what I hear of her, that she is growing up a charming girl. Sarah, I think, is really

* She was then 15.

lovelier and more engaging than ever. I do think 1868. that she and the other Sarah (Hervey) are two girls quite incomparable in their different ways.

Much love to dearest Cissy,

Ever your affectionate brother,

CHARLES J. F. BUNBURY.

JOURNAL.

Saturday, 24th October.

Excessively wet. Finished reading the third volume of Kinglake's "Invasion of the Crimea." This volume of 486 pages, contains the history of not quite four weeks:—from the night following the battle of the Alma (September 20th), to the end of the unsuccessful canonade on the 17th of October. It is certainly written with excessive—I think unreasonable diffuseness; there are many seemingly unnecessary repetitions, and the author seems to have a predilection for telling every fact in a circuitous way, rather than straightforward. The most interesting part of the volume, to my mind, are those which he has derived from Russian sources. The characters of Korniloff and Todleben, the two heroic and devoted men who (though themselves almost without hope) undertook the defence of Sebastopol when it was left to its own resources by Prince Mentschikoff, are particularly interesting and striking. The last hours of Korniloff, and his death in the bombardment on the 17th October, are extremely well told.

1868.   Kinglake is of opinion that in the course of this month, the Allies neglected or failed to seize several favourable opportunities of taking or destroying Sebastopol and ending the war. First, immediately after the battle of the Alma, when, he holds, they would have been able by a bold advance and an immediate assault to take the Star Fort,* to gain possession of the north side, and to have Sebastopol and its harbour at their mercy. Secondly, when after the flank march they arrived on the heights on the south side of Sebastopol. He maintains that at that time, the defences of the town were so imperfect and the garrison so scanty, that a vigorous assault by the Allies (without losing time in landing their siege guns) must probably have been successful. In this, again, he is supported by the high authority of Todleben. Again at the close of the canonade of the 17th October, when, although the French fire had been silenced, the Redan had been almost entirely ruined by the English fire: he holds that an immediate assault might have been successful, but this seems rather doubtful. Kinglake's main conclusion in this volume, seems to be much the same as that of Macdougall—that the *divided* command was alone sufficient to account for all the failures of the siege of Sebastopol: but he goes further, and probably too far, in maintaining that in all the

* In this opinion he seems to be borne out by Todleben, who seems to represent the Star Fort, as at that time insufficiently manned and of little strength. But I find that William Napier is of opinion that the possession of the Star Fort and the north side, would not have given us the command over Sebastopol and the Russian Fleet.—(C. J. F. B.)

instances given in this volume, it was the opposition 1868.
of the French commanders which caused opportunities to be neglected ; several maps and plans of
Sebastopol in this volume are very instructive and
useful. Fanny and the rest went to a concert at
Bury.

———

Monday, 26th October.

Mr. and Mrs. Lombe arrived with their daughter
Julia, also Mr. and Mrs. Bowyer, also Major Tyrell,
Captain Eyre and Mr. Lott. The Suttons and Mr.
Percy Smith dined with us.

———

Tuesday, 27th October.

A very fine day. I showed Mrs. Lombe the
arboretum and Fern houses. Sarah and Kate
Hervey, Lady Cullum, Captain and Miss Horton
and Patrick Blake dined with us. Lady Cullum
very amusing.

———

Wednesday, 28th October.

We walked about the grounds with Mrs. Lombe.
Another beautiful day. Mr. and Mrs. Abraham and
Mr. and Miss Bevan dined with us.

———

October 29th. Thursday.

Mr. and Mrs. Lombe and Julia left us. I retain
my old regard for them. The gale of wind on
Saturday the 24th, blew down a large "Ivy Tree,"
(a dead trunk covered with a luxuriant mass of Ivy)
in the arboretum. The large red maple, Acer
rubrum, in the outer arboretum (Sorcerer's

1868. paddock) has now a most beautiful appearance; the whole of its foliage variegated with the most exquisite tints of red and yellow, shading off from crimson and scarlet through bright orange into amber colour, and here and there a little green still mixed with the rest.

———————

Saturday, 31st October.

Augusta and Frederick Freeman and the Bowyers went away.

═══════

LETTER.

Barton,
November, 2nd, 1868.

My Dear Katharine,

I have long felt rather uneasy in my conscience about my not writing to you ; but for some time I must confess I was idle and dissipated, and for the last fortnight I have been hindered by an obstinate worrying cold, which I have not yet been able to shake off, and which has interfered both with my enjoyments and my usefulness. I believe Fanny has sent you from time to time her journals, giving you some idea of the pleasant company we have had ; and I am afraid I have not much of interest to tell you.

With the exception of a few very cold days about a fortnight ago (at the beginning of which I caught this cold at Church), the autumn has been and still is a very fine one, and the beauty of colouring of

the foliage has been beyond what I remember to 1868.
have ever seen before. I can hardly imagine that
even the North-American woods can show more
glorious colouring. The Liquidambers and the
Virginian Creepers have been most gorgeous; and
there is a Red Maple, Acer rubrum, which is, or was
a few days ago, a mass of the most brilliant red,
orange and yellow, shading off in beautiful gradations from scarlet to amber colour.

The trees and other plants which have been most
gloriously coloured are all American, but they are
well supported by our Beeches and Sycamores. I
have been more than ever struck by the skill with
which my father arranged his trees, with a view to
the effects of the vernal and autumnal colouring.
The produce of *fruit* (in the botanical sense) is quite
extraordinary; the Catalpa and the black Walnut
are perfectly loaded; I have got ripe fruit of the
Æsculus macrostachya (which I never saw before)
of Æsculus Indica, of the Paliurus and various
others.

Edward is indefatigable in his canvass for Bury,
and is in very good spirits; I hope he will be
successful. Except in this case and a few others
where I am personally interested in the candidates,
I do not care much about the elections, for I do
not like either of the great parties. I hope I shall
be able to shake off my cold before the election
time, when as Sheriff of the County, I shall have to
appear on the hustings.

I am going on by little and little with Darwin's
"Variation." I am in continual amazement at the

1868. quantity of facts that he has collected and arranged; and now that I have got into the botanical department, I shall be better able to judge of his reasoning upon them than I can in the case of the birds and beasts.

I am also reading Heer's "Flora Fossilis Arctica," which I like very much, the general or introductory part, that is. It is really very wonderful that there should be such clear and irresistible proofs of a rich forest vegetation having existed where now there is nothing but snow and ice. Also I am still going on at my leisure with the Odyssey.

I am very sorry to say that poor Scott is again in anxiety about his family; one of his younger children has been attacked by scarlet fever, and I fear it may spread to others. I do hope they will get well over it.

Dear Minnie and Sarah leave us to-morrow,* which I am very sorry for ; it is only for a couple of days, indeed, but I am afraid they will stay but a short time longer with us. Mr. and Mrs. Walrond are just arrived ; Mrs. Walrond is Mrs. Kingsley's niece, with whom I was so much charmed in London.

Ever your very affectionate brother,

CHARLES J. F. BUNBURY.

---

* To go to Riddlesworth.

JOURNAL.

Mrs. and Miss Eyre came to luncheon. Mr. and 1868. Mrs. Walrond and Mr. Clark* arrived. Mrs. Walrond charming. Accounts in the newspapers of a slight shock of earthquake felt extensively in the West of England and in Wales.

Weather very stormy. Mr. Augustus Tollemache and Mr. Percy Smith came to luncheon. I read some more of Kinglake. Edward brought from Bury the bad news of an opposition for West Suffolk. Lady Cullum dined with us.

A beautiful day. News arrived of the success of General Grant in the Presidential election in the United States. I walked through the arboretum and gardens with Mrs. Walrond. She was delightful.

A sharp white frost in the morning. Mr. and Mrs. Walrond and Mr. Eddis went away. I was very sorry to part with the Walronds so soon. Mrs. Walrond is quite as charming as I thought her the first time: remarkably handsome, with a sweet voice, most winning manners, much knowledge and

* Public Orator at Cambridge

1868. accomplishment, enthusiasm, and if I am not much
mistaken, a truly fine and noble mind.   She is
worthy to be Mrs. Kingsley's niece.   I had an
electioneering visit from Mr. Lamport, the new
Liberal candidate for West Suffolk, accompanied
and introduced by Mr. Manning Prentice.   I have
an agreeable impression of the former.

Saturday, 7th November.

A long talk on election business with the Under
Sheriff.   Dear Minnie and Sarah returned from
Riddlesworth.   Susan and Joanna arrived, also Mr.
Strutt.*

Monday, 9th November.

Dear Minnie and Sarah went away—very much
regretted, also Mr. Strutt and Clement.

====

LETTER.

Barton,
November 10th, 1868.

My Dear Henry,
        Many thanks for your letter.   Mr.
Lamport's coming forward was quite a surprise to
me, and a disagreeable surprise, for of course it
would have saved me, as Sheriff, much trouble if
there had been no contest for West Suffolk.   At
first, therefore, I was disposed to look unfavourably
on this stranger: but he called upon me, and I
confess that my impression of him from the inter-

* Afterwards Lord Rayleigh.

view was favourable. He is, I think, unmistakably a gentleman, and Edward thinks so quite as decidedly as I do. His manner is modest and quiet: he professes no extravagant opinions, and he is connected with some men, whom I hear well spoken of.

I quite agree with you, that it would have been much better if some gentleman of a good county family and of known character and position in the county, had come forward to represent the Whig party, but none such could be found. The "Liberals" applied first to me, and in a very handsome and flattering manner, expressing great respect and esteem for the family: but (independently of my official trammels as Sheriff) many reasons had made me determine to have nothing to do with parliamentary life. Then they asked Edward, but he also declined: so did Sir Charles Rowley, both for himself and his son.

Maitland Wilson has become an avowed Conservative,—so is Tyrell: and it seems that no member of the house of Fitz Roy was available. Whig gentlemen of high standing and character are not as plentiful in Suffolk as partridges. So, as the Liberals were determined to have a fight for the division, they have done probably the best they could. As far as I can at present see, I think that I shall give Mr. Lamport a vote: but, I speak conditionally, for something *may* come out (as to his opinions, I mean) between this and the day of election, which may alter my views: I think I shall give a vote also to Lord Augustus Hervey, both

1868. from friendship for his family, and because I think it a great point to oust the old Tory who is Lord Augustus's colleague. Not that I think it probable that the said Tory will be ousted, but it is something to *try*.

I have gone at length into this matter, because I am much gratified by your wish to act with me in it, and I wish to explain to you how things appear to me. I hardly suppose you will think it worth while to come all this long way to vote, but if you should, you know how very glad we shall be to see you.

The polling for West Suffolk will be on Saturday the 21st; the *declaration* of the Poll on Monday, the 23rd; and on that same Monday, we must set off for Stutton, for the nomination at Ipswich on the 24th.

Edward is indefatigable in canvassing, and is in very good spirits; hopeful, though not confident of success. Except for him and a few others, for whom I have a personal regard, I feel no great interest in the elections.

The state of politics appears to me thoroughly unsatisfactory, and I have little hopes from any change. I utterly distrust Disraeli, but I should have no satisfaction in seeing a Cabinet formed under the influence of Democrats and Socialists.

Fanny is very eager about the Bury election; if I were standing, I think she would be in a frightful state of mind.

I hope we shall see you in London, as we mean to go up as soon as the elections are over—either on the 28th or 30th. With very much love to dear Cissy and the darling children,

I am ever your very affectionate brother,
CHARLES J. F. BUNBURY.

JOURNAL.

Monday, November 9th.

Dear Minnie and Sarah went away, very much regretted. Also Mr. Strutt and Clement. Lady Cullum dined with us.

Wednesday, 11th November.

Finished reading vol. iv. of Kinglake's " Invasion of the Crimea." This volume is almost entirely occupied with the singular events of the famous 25th October. The advance of Liprandi's corps towards Balaclava ; the capture of the field-works occupied by the Turks ; the repulse of a part of the Russian cavalry by the 93rd Highlanders ; the splendid charge of the heavy cavalry under General Scarlett, and the ever memorable charge of the light cavalry brigade under Lord Cardigan.

The battle of Balaclava, as Kinglake says, consisted in fact of four separate combats* with very little connection between them.

* Kinglake indeed says five, but I do not see how he makes this out. Yes, —I suppose he distinguishes Dr. Allenville's charge with a part of the French cavalry as one of the five acts.—(C. J. F. B.)

1868.   The very elaborate portraits of Lord Lucan and
Lord Cardigan with which Kinglake introduces his
narrative of these operations, are drawn with
extraordinary brilliancy and ability. Of their justice
I cannot judge, but as pieces of historical character
painting, I think they are most powerful; most
severe, but with a delicate discriminating touch;
not virulent or abusive.

Kinglake describes, with great fulness and
copiousness of detail, the heavy cavalry charge, and
takes great pains to show its importance, which as
he perceives has been unduly eclipsed by the more
startling exploit of Lord Cardigan's brigade.

General Scarlett's charge ought certainly never to
be forgotten, for it was a splendid operation of war,
perfectly well planned and executed, and thoroughly
successful against enormous odds. The light
cavalry charge is also related in very great detail,
yet extremely well, without tediousness and without
loosing the general effect, amidst the multitude of
particulars.

Altogether this volume, though it is almost wholly
filled with the events of one day, is far from tedious,
and is much more interesting than the third volume.
It seems clear from Kinglake's statements, which
appear to be very careful, that the blame of the
mistake which threw away our light cavalry, must
rest between Lord Lucan and Captain Nolan, the
greater share of course falling on Lord Lucan as the
superior officer. But I cannot agree with Kinglake
that Lord Raglan's order was clear and unmistakable in its meaning.

There seems to have been a strange hesitation 1868. and want of decision and vigour, in all the proceedings of the Russians on this 25th of October. After taking the earthworks held by the Turks, and laying open the weakness of our position at Balaclava, they stopped short, and remained for hours without attempting anything more. Their demonstration against the 93rd was a mere demonstration, bloodless and harmless. When their great body of cavalry at last advanced, it was without any evident object ; and when they found our heavy cavalry before them, instead of taking advantage of the slope and charging home, they halted to receive the charge.

Thursday, 12th November.

Dear Susan and Joanna went away. We were very sorry to part with them.

Went to look at the new North Lodge.

We dined with the Wilsons. Met the Maharajah, Mr. and Mrs Fuller Maitland, Sir Richard Kindersley and others. Sir Richard Kindersley says that to try to get into Parliament is—*standing* with the chance of *sitting* and the certainty of *lying*.

Friday, 13th November.

News of the birth of Mrs. Walrond's baby.

The new dog arrived—very handsome. We called him Boy.

Sunday, 15th November.

Looked over Boswell's "Johnson." Read the remainder of Stanley's Commentary, on 1. Corinthians, chapter 15. This Commentary is like everything else of Stanley's writing, clear, sensible and full of the best feelings ; and he does something to clear up the obscurities of this famous chapter. But he does not satisfy me as to what has always appeared to me an imperfection in St. Paul's reasoning; namely, the bringing the miraculous resurrection of our Saviour as an evidence of the *general* resurrection of mankind.

Monday, 16th November.

Much pleased at the news that our good friend Dr. Tait, the Bishop of London, is made Archbishop of Canterbury. I do not think a better man could have been chosen.

Very damp and foggy, I did not go out. Fanny went with Edward into Bury, to his nomination, and returned to luncheon.

Tuesday, 17th November.

The Bury election. Edward unsuccessful, I am very sorry to say. The numbers polled were—

For Greene ..	714.
Hardcastle	793.
Bunbury ..	593.

Fanny and I went to Bury in the afternoon, and stayed till after the close of the poll. Saw Mrs. Abraham and Patrick Blake.

The nomination for the Western Division of Suffolk at Bury. I presided as Sheriff, and opened the proceedings. Major Parker was proposed by Henry Waddington and seconded by Mr. Hitchcock of Lavenham. Lord Augustus Hervey proposed by Mr. William Bevan and seconded by Mr. Biddell. Mr. Lamport proposed by Mr. Pettiward, seconded by Mr. Manning Prentice. The proceedings took up about two hours, the show of hands was for the old members. The behaviour of the crowd was, all things considered, tolerably orderly.

LETTER.

Barton,
November 19th, '68.

My dear Katharine,

I this day send off by post your two packets of papers on Ferns, which I trust will reach you safely. I have gone through them all, and found scarcely anything to correct, and little to add, as I have no *special* knowledge of the botany of those countries,—no knowledge otherwise than from books. Your work is a very laborious one, and I am very glad you are getting on so well with it: very glad, too, whenever I can help you.

Our *Liberal* friends are strangely unlucky.—I mean those Liberal candidates in whom we felt a

1868. personal interest, but I am most especially sorry on account of Henry Bruce, whom I look upon as a very valuable man.

I am much disappointed about Edward, but he had not latterly been feeling very sanguine of success, and he is very good-humoured about his defeat.

I am thankful that I have got through the first stage of my work—the Nomination for West Suffolk, without catching cold; I must pray that I may be equally fortunate next week. I have very good hope of Shafto Adair's success in the Eastern Division.

Fanny is gone to Mildenhall for the day, to see schools and curates, &c. Henry is coming to us to-morrow—coming all the way from Wales to vote for the Liberal Candidate for West Suffolk.

I am delighted at the appointment to the Archbishopric of Canterbury,

Ever your very affectionate brother,
CHARLES J. F. BUNBURY.

JOURNAL.

Friday, 20th November.

Fine day. Walked round the grounds with Fanny and her new dog. Henry arrived.

Saturday, 21st November.

I went into Bury with Henry, and we voted. Mr. Lamport the Liberal Candidate lost, as was to be

expected, but he made a good fight. The numbers
at the close of the poll were:—

For Parker	..	2500.
Hervey	..	2389.
Lamport	..	1704.

The Arthur Herveys came to afternoon tea with
us—as delightful as ever.

Monday, 23rd November.

In the morning the declaration of the Poll for
West Suffolk—not many present. Edward and
Henry left us. We left home at 3 p. m. and
travelled post to Stutton to our dear friends the
Mills's, who received us with their never-failing
cordiality. We arrived at 7.20.

Tuesday, 24th November.

We breakfasted early and went into Ipswich to
the Nomination for East Suffolk. The candidates
were the Hon. Mr. Henniker and Mr. Corrance on
the one side (the Conservative) : Shafto Adair and
Mr. Sutton Western on the other. The proceedings lasted three hours and-a-half and were very
fatiguing to me as I was standing all the time, and
I was excessively crowded and pressed upon. The
multitude was much greater than at Bury, and very
noisy, but good-humoured and not at all violent.
Much the best of the speeches on this occasion was
that of Milner Gibson who proposed Mr. Western.
Adair's speech was good, and though long, not
tedious. Austin (who proposed Adair) clever, but
much too long.

Wednesday, 25th November.

A quiet day at Stutton. I took a short walk with
John Hervey. Read part of the Memoir of Hugh
Elliot. Mr., Mrs. and Miss Rimmer and Miss
Kirkpatrick dined at Stutton.

Thursday, 26th November.

The principal Ilex trees at Stutton are five in
number, growing in a group so that their branches
intermix, and they form a complete and compact
dome of foliage, the outermost branches everywhere
sweeping the ground. The circumference of this
dome, as I have measured it to-day by stepping
round the edge where the ends of the branches rest
on the ground, is upwards of 100 yards.

Mr. and Mrs. Henry Keene, Mr. Zincke, Mr.
and Mrs. Nicholl and Mr. Pigot came to dinner.

Mr. Zincke spent last winter in America. He
told me that one of the things that struck him in
that country (where he travelled extensively, even
to the rocky mountains) was the general absence
of all *green* grass in winter: all the grasses seem to
die down to the root at that season, the leaves as
well as stalks of the season die down (as in our
common reed), and no green leaf remains. He
had also been struck by the want of *large* trees in
the native forests, and by the general similarity of
the appearance of those forests to our English
woods. Of course this must be taken with much
allowance, as he was there in winter, when all the
deciduous trees were leafless, and he could not see

the differences between (for instance) the different species of Oak, but the similarity of the *general* appearance of the woods under these circumstances to ours, was striking to him: and this agrees with Agassiz's remark on the wide differences of the *Brazilian* scenery from that of North America.

Mr. Zincke was not in America at the season for seeing the beautiful autumnal colouring of the foliage. What he remarked as most *exotic* in the way of vegetation, were the "cane brakes," (thickets of a slender species of bamboo) in the valley of the Mississippi.

—————

Friday, 27th November.

Received a very agreeable letter from Edward Romilly about my book.*

Read Gladstone's "Autobiography." Mr. Mills showed us his collection of views of Suffolk Country houses.

—————

Saturday, 28th November.

Received the bad news of Adair's defeat. We drove to Shotley to see John Hervey.

—————

Sunday, 29th November.

A beautiful morning. We went to morning church, where Mr. Mills read the service beautifully.

—————

Monday, 30th November.

We left our kind friends at Stutton and drove to

—————

* His Father's Life

1868. Ipswich. The declaration of the poll for East
Suffolk,—long speeches, great crowd and excitement. We had luncheon at the White Horse and
posted home—arrived safe and well, thank God.

———

Tuesday, 1st December.

During the week we spent at Stutton (while the
election was going on), I read the Memoir of Hugh
Elliot, (brother of the first Lord Minto), by Lady
Minto: very agreeable reading. Read also Gladstone's remarkable pamphlet: " A Chapter of Autobiography." This interested me much: it is written,
as it appears to me, with a noble candour and
true dignity of feeling, and is indeed a very striking
record of the changes through which his mind
has passed with relation to one subject, that of the
Church. While vindicating himself from changes
of interested factions or hasty inconsistency, he
does not at all deny the incompatibility of his
present opinions on the Irish Church, with those
which he held at the beginning of his political
career. He recapitulates the views on Church
matters with which he started in public life
as expounded in his famous book on " Church and
State;" he fairly admits that he *now* considers
these views as entirely untenable, but he points out
what were the causes which led him to form these
opinions, and what were the facts which at that
time seemed to justify them. In this part he draws
a very forcible picture, (and I daresay in the main a
very true one), of the change in the spirit of the
Church of England and the general character of the

clergy. He then proceeds to trace the effect on 1868.
his own political conduct, of the discovery that his
theory of Church and State was untenable.

In conclusion he states the reasons why he now
wishes to disestablish and disendow the Irish
Church, and here I must say he affords a justifica-
tion to those who say that he (Gladstone) views the
disestablishment of the English Church also as
merely a question of time.

The style of this pamphlet is dignified, but somewhat involved and cumbrous, especially in the
earlier part.

————

Wednesday, 2nd December.

Edward tells me that the elephants which are
represented on several of the later Greek and
Græco-Punic coins are, except in one instance,
distinctly of the African species, clearly marked as
such by their enormous ears. The exception is in
the case of Seleucus the 1st, who is recorded to have
made or sent an expedition to India; as on one of his
coins Indian Elephants are represented drawing a
car.

————

Thursday, 3rd December.

The surprising news arrived of the resignation of
the Disraeli Ministry.

————

Friday, 4th December.

Most part of the day spent (or rather wasted) at a
meeting at Bury to consult about a visit of the
Archæological Institute.

————

Monday, 7th December.

A furious gale in the night, morning very fine.
Much business with Scott.   We walked round the
grounds, and went on with our list of trees for
labelling.   I went on with my herbarium catalogue.
Finished reading Nassau Senior's " Journals and
Conversations and Essays relating to Ireland,"
that is, as much as I intend to read, being the
Journals and Conversations which are comprised
in the last 50 pages of the first volume and the
whole of the second.

It is a book which has interested and impressed
me very much but I have not time just now to write
down what I have to say of it.   A melancholy and
disheartening picture of the state of Ireland.

———— ————

Wednesday, 9th December.

Up to London.   We arrived all right, thank God.
A visit from dear Mary.

———— ————

Thursday, 10th December.

A fine day, extremely mild.

I called on the Miss Boileaus (Minnie and
Caroline), had a pleasant talk.   Called on the
Adairs.   We drank tea with Charles and Mary, and
met Katharine.

Lyell told me that he met at the Geological Club,
a man lately come from Arabia who had visited the
Turquoise mines there* who told him that in the

---

* From a subsequent conversation of Lyell's I find that my note on the
subject of the Turquoise mines in Arabia, was pretty correct.   The mines are
believed to have been worked 4000 years ago, no doubt by the Egyptians, or
at least on account of the Kings of Egypt.

recent re-opening of those mines (as I undertand) 1868.
many of the tools used in the ancient working had
been discovered, and these were all of flint, and very
similar to the flint hatchets of Europe. The flints
were obtained from the *nummulite* limestone
(Eocene) of that country, in which it occurred in
much the same circumstances as in the European
chalk.

In a paper by Boyd Dawkins in the November
number of the *Quarterly Journal* of the Geological
Society, I found information which I had long
wished for about the native country of the Fallow
Deer. Dawkins (and he is a high authority)
is satisfied that it is not an aboriginal native
of any part of Northern or Middle Europe,
but of the countries bordering the Mediterranean,
especially Northern Africa and Asia Minor. In
Britain, he says, its remains have not been found
in any deposits older than the historic times; and
he concludes that there is every probability that it
was introduced by the Romans (as I believe was
also the case with the Pheasant and the Chesnut
tree).

————

December 11th.

We went out in the brougham—paid visits. Saw
Miss Phillips, Mrs. Young and the Edward Romillys
—very pleasant. Clement arrived.

Miss Phillips told me that her brother, when canvassing for one of the divisions of Cheshire, saw
much of the higher class of Dissenters (what he
calls the aristocracy of dissent)—and the general

1868. result of what they told him was this : that the
Dissenters are not now hostile (politically) to the
Church establishment in England ; not disposed to
aim at its overthrow : but they looked on the Church
establishment in Ireland as an injustice and a
scandal, and if the measures of the new Government shall be baffled through the opposition of the
English Church, they (the Dissenters) will have to
take up a position of declared hostility to the
Church of England.*

December 12th.

Went with Fanny to some shops. Called on
Lady Rich. Read some more of Darwin on
Variation.

Monday, 14th December.

A visit from Norah Bruce, and a long and delightful talk with her. I think she is one of the most
admirable women I know. I said to her with great
truth that I do not so much congratulate her or her
husband on his appointment to be Home Secretary,
as rejoice that the country has secured his services
in such a supremely important post. She told
me that it is not yet known what borough he will be
returned for ; but as his appearance in the House of
Commons will not be required before February, he
has time to look about him. We agreed that the
question of the Irish Church would be found full of

---

* *March 10th*, 1871. — The Irish Church has been overthrown, and the
political Dissenters *have* nevertheless taken up a position of declared hostility
to the Church of England.

practical difficulties, as soon as the first general 1868.
declaration of principles was got over,—she told me
that when the new Ministers went down to Windsor to
kiss hands, the Queen showed especial courtesy and
attention to Mr. Bright. She caused it to be notified
to him that he would not be required to dress specially
for the occasion, to kneel, to kiss hands, or to go
through any ceremony of which he did not entirely
approve. He declared that he would on no account
omit to kiss her hand ; and he did kiss it, but did
not kneel. He is not to wear the Ministerial uniform. Norah said it is thought that Mr. Lowe
is likely to be the most explosive element in the new
Cabinet ; more than Bright. She said her husband
is very apprehensive lest Mr. Foster should be
unseated on petition ; he would be a great loss to
the Government, and especially to Henry Bruce,
with whom he has worked much in unison on
educational questions.

Norah has, like me, been reading Senior's Conversations on Ireland, and agrees with me that it is
an exceedingly interesting book, but that it leaves
an exceedingly melancholy impression of the state
of that country. One thing she thought particularly disheartening, is the determined bigotry and
spirit of proselytism infecting the most educated
Protestants ; not even the excellent Archbishop
Whately appearing free from it.

I called on Sir John Bell and Lady Smith—both
pleasant. Read to Fanny Longfellow's new book,
"The New England Tragedies."

Tuesday, December 15th.

We dined with the Charles Lyells. Met Sir
Louis and Lady Mallet, Mr. and Mrs. J. Evans,
Julia Stirling, John Moore, Lady Bell and others in
the evening.

—————

Wednesday, December 16th.

Henry Bruce and Norah dined with us. He told
me that it was quite true that Bright was very
unwilling to take office, and was with much
difficulty persuaded to do so. This was not on
account of any Quakerish scruples, but because he
felt himself unfit for the hard work of administration.
At last he accepted the office, which of all those
included in the Cabinet has (Bruce says) the least
work and the least pay. Bruce spoke of the
difficult and painful questions which sometimes
occur to a Home Secretary, in relation to the
punishment of death. He says he has laid it down
as a rule, never to consent to receive a deputation
relative to the case of any prisoner under sentence
of death: he said, it might be supposed to be a safe
rule that in every case of *deliberate* murder the
sentence of the law should be carried out: but
there do occur cases, now and then, in which even
this rule has been hard to enforce.

Bruce talked with great delight of the holiday
time which he spent last autumn in the Island of
Harris, the outermost of the Hebrides. He spoke
quite with enthusiasm of the interesting wildness of
the country, the grand rocks, the magnificent surge
that rolls in on the coast, the extraordinary
appearances of the storms, the eagles sailing about
the towering rocks.

I saw the Youngs and Helen Richardson. We dined at the Henry Lyells:—met the Benthams, Wedgwoods, and Charles Mallets.

Charles Lyell tells me he has in his possession some of the stone implements (so-called hatchets and arrow heads) made of *Quartzite* (granula quartz), and found in the Deccan in India; the localities and circumstances are described by Mr. R. B. Foote in the *Quarterly Journal* of the Geological Society. They are of the true rude *Palæolithic* type, and not unlike the flint implements found in Europe. They are supposed to have been made and used by some of the early native tribes which preceded the Hindoos in that country, and which are believed to have been allied to the aborigines of Australia.

Minnie Boileau called—very agreeable. She went out driving with us. I bought an opal ring for Fanny. A visit from Mrs. Walrond and two of her children—she was as charming as ever.

I read prayers with Fanny, and Kingsley's fine Sermon on the "Humanity of God." Visits from Charles and Mary and the Bowyers. Began to read to Fanny "The Spanish Gipsy."

1868.   Monday, 21st December.

Went to see the Napier children at Wimbledon.

_____

Tuesday, 22nd December.

William Napier came to luncheon with us: just returned from Malaga.—We were right glad to see him.   I went with Fanny to Hennell's, changed the opal ring for her, got an emerald ring instead. Visits from Mary, Katharine and Edward.

_____

Wednesday, 23rd December.

We returned home.   The weather very wet and dark, not cold, in fact the mildness of the temperature all this month has been remarkable. We returned by the line of rail by Marks Tey and Sudbury.   Found all well at home, thank God.

_____

Thursday, 24th December.

Stormy weather, and some of the storms accompanied by loud claps of thunder, very unusual at this time of year.

Read a little of Raleigh's Preface to his "History of the World."   In this history he brings together a most tremendous collection of the crimes of kings: such as might satisfy the fiercest Republican. What he says of Henry VIII. (to whom he assigns a conspicuous place in this list of wicked kings), is worthy of notice, as showing that the commonly received opinion of Henry's character, if erroneous, is as Froude contends, not a modern error.

Christmas Day. We went to morning Church.

I read the remainder of Raleigh's Preface to his "History of the World." This is partly a moral discourse, (for the most part more true than new), and partly a rather lengthy disquisition on the origin of the physical universe in opposition to atheism. I hope to find something more interesting as I go on with the work. The preface is not dated, but at the end of it he tells us that "it was for the service of that inestimable "Prince Henry the successive hope and one of the "greatest of the Christian world, that I undertook "this work."

Went on reading "The Spanish Gipsy" to Fanny. Read several papers in the *Idler*.

I went out and visited the Ferns and walked round the arboretum.

I will now put down a few remarks on Senior's "Journals and Conversations in Ireland," (See December 7th). These conversations are, as I have said before, uncommonly interesting: they have the appearance as far as one can judge, of being honestly and fairly reported, and they convey the sentiments and experiences of a considerable variety of persons:—of Lord Monteagle, of Archbishop Whately, of several gentlemen who had been employed as agents for great estates in different parts of Ireland, of Mr. de Vere, a zealous

1868. Papist, of some ladies who were zealous Protestant propagandists, and of various others. The general impression it has left on me is melancholy in the extreme, almost to hopelessness.

I hardly know whether the religious or the agrarian difficulties appear the more deplorable. It may be true, as some of the speakers contend, that the murders of landlords and agents in Ireland evince less of actual moral depravity in the people than the miscellaneous murders in England. But, the Irish species of murder being aimed directly and especially against all who attempt to improve their estates: being therefore essentially hostile to all improvement of the country: being based upon an altogether barbarous and semi-savage theory of landed tenure: being consciously and systematically hostile to the law of the land; is far more dangerous and more imcompatible with any orderly state of society than English species. After reading the various instances given in these conversations of murders and attempted murders of landlords and agents, I wonder more than ever, how anyone can ever endure to live on his estate in that country. Nor do I see what remedy there can be for such a state of things. As for the religious question, there also the evils appear enormous and deeply rooted; it certainly appears to me very doubtful whether "Disestablishment" would do much towards softening the embittered rancour between the religions, diminishing the spirit of proselytism, or correcting the virulence of bigotry. The picture drawn of the Popish priesthood of Ireland in

these conversations is most deplorable : — excessively ignorant, bigoted and narrow-minded, in
proportion, bitterly disaffected to the Government, sharing the passions and prejudices
of the lowest classes, encouraging them in
their false notions of property and in their
enmity to law. But on the other hand, in the
members of the established church, there appears
to be much that is unpleasing: much of bigotry and
narrowness of view, a habit of looking on those of
the other religion with an undiscriminating aversion
and contempt, and a restless spirit of proselytism
which exasperates the opposite party and tends to
aggravate their faults.

One conclusion to which I am led by Senior's
book, and it is a very disheartening one, is this, that
since the famine, the physical condition of the
Irish people in general has been very materially
improved: but at the same time the state of feeling
and the disaffection to the Government, the
animosity between those of the opposite religions,
the ill-will of the peasants towards their landlords,
all have been worse than before. The moral evils
are aggravated, while the physical evils are much
alleviated.

As for the religious difficulty, it is, I think, quite
clear from these conversations the best thing to do,
if it were still possible, would be that the State
should endow the Roman Catholic priesthood.
Whately was particularly earnest and emphatic on
this point, and what he told Senior of Lord Grenville's opinion, is very striking. Lord Grenville was

1868. always a zealous supporter of "Catholic Emanci-
pation," and when Wellington and Peel's measure
to that effect was carried in 1829, some of his
friends wondered that he did not seem to rejoice.
"You do not pay the priests," he said, "and therefore you are doing more harm than good by giving
them mouth-pieces in parliament. But it seems to
be assumed that it is now too late to adopt this
course.

Read the Odyssey, book 13, v. 97-184. The
arrival of Ulysses in Ithaca and the anger of Neptune against the Phæacians. There is something
touching in the description of the deep repose, the
happy undisturbed sleep which Ulysses, the man of
so many toils and sorrows enjoys ou the voyage,
and in which he is deposited on the shore of his
own country.

I finished the first volume of Darwin on "The
Variation of Animals and Plants in Domestication."

————

Monday, 28th December.

Furious storms of wind and rain.

I see in the *Gardeners' Chronicle*, the death of the
great botanist and traveller Von Martius. He died
at Munich on the 13th, at the age of 75.

Read the Odyssey, book 13, v. 185-310. Much
of the *wisdom* of Ulysses seems to have consisted in
a readiness to tell any amount of lies:—he meets
Minerva, disguised, and taking her for a stranger,
forthwith prepares to tell her a long story about
himself, which is entirely false from beginning to
end: and this without any apparent necessity. It

would seem as if the Greeks of Homer's time had 1868.
much about the same esteem for truth as the Greeks
of the present day.

Tuesday, 29th December.

Excessively dark and wet. I could hardly see to
write in the morning.

Read the Odyssey, book 13, v. 311-440, the end
of that book. Minerva promises Ulysses her
assistance, in redeeming his kingdom and
possessions, and begins by giving him the outward
appearance of a beggar.

Wednesday, 30th December.

Rain, sleet and a little snow. Barton estate
accounts. Received an agreeable letter from
Kingsley. A visit from Maitland Wilson and Tom
Thornhill. Read a little of Raleigh's "History of
the World." In book 2, ch. 4, section 13, he mentions
incidentally, that there was in his time more of
robbery with violence in England "than in any
region of the world among Christians," and he
attributes it to a want of strictness and severity in
executing the law.

I have gone on from time to time reading portions
of Raleigh's "History of the World," but as far as
I have yet gone, I must confess I am disappointed.
He is excessively lengthy, running continually into
dissertations of little interest on such questions as
the true locality of the terrestial Paradise: and
whether the tree of knowledge was the Ficus Indica;
and when he comes to the law of Moses, he launches

1868. out into a long discussion on the nature of law in general. It is not difficult to believe that he wrote this book during his imprisonment; he certainly wrote it when he had plenty of leisure.

I finished reading to Fanny " The Spanish Gypsy" by " George Eliot" (Mrs. Lewes)—a very beautiful poem, though a most improbable story. The two principal characters (grand and terrible characters)—Zarca the gypsy chieftain, the fanatic for his nation; and Isidor, the Inquisitor, the fanatic for his Church—are finely contrasted. The death scene of Zarca is grand.

———

December 31st.

Fanny ill with bilious sick headache. I went alone to the Arthur Herveys—a most agreeable family dinner, afterwards a large party. Charades acted capitally by the young people of the family, together with Lady Mary, Lord John and Lord Francis. I slept at Ickworth.